THE AGE OF CONSEQUENCES

THE AGE OF CONSEQUENCES

A CHRONICLE OF CONCERN AND HOPE

COURTNEY WHITE

INTRODUCTION BY WENDELL BERRY

COUNTERPOINT | BERKELEY, CALIFORNIA

Library of Congress Cataloging-in-Publication Data Is Available

ISBN 978-1-61902-454-0

Cover design by Gerilyn Attebery
Interior Design by Elyse Strongin, Neuwirth & Associates, Inc.

Counterpoint Press
2560 Ninth Street, Suite 318
Berkeley, CA 94710
www.counterpointpress.com

Printed in the United States of America
Distributed by Publishers Group West

10 9 8 7 6 5 4 3 2 1

for Sterling and Olivia

CONTENTS

INTRODUCTION
BY WENDELL BERRY

IN 1997, WHEN Courtney White and his allies started the Quivira Coalition, conservation in the United States, and most conspicuously in the west, was in serious failure. Wilderness conservation was more or less all right, and the conservation organizations would dependably rise up and shout in emergencies such as oil spills, but the landscapes of farming, ranching, and commercial forestry were often seriously abused, as often they still are, without much notice by anybody. The official forces of forestry and agriculture promoted "best management practices" that were, and still are, "best" by standards unrevealed. These practices were, and are, mostly unquestioned by a public trained in submission to the opinions of experts.

The land uses that *had* caught the attention of the conservationists were mainly ranching and logging on the public lands of the west. In the grazing lands of the southwest, where Courtney lived and had worked for one of the established conservation organizations, the ranchers and conservationists were divided by a political and rhetorical hostility that both sides had worked diligently to perfect. The conservationists were assuming, to the extent almost of misanthropy, that there was a contradiction between any economic use of the land and the land's natural character, a contradiction that was absolute and unamendable: just get the humans and their needs off the land; then the land will return to nature and be well. The ranchers, cherishing their belief in their independence, stood on their rights and wished to be rid of the conservationists, whom they saw as ignorant city people meddling in other people's business.

Courtney had spent enough time on the conservationists' side of this division, and he had enough sense, to see that this feud was useless and wasteful. Nothing good was ever going to come out of it because nothing good was in it. The Quivira Coalition came from that realization. The Coalition placed itself in the middle between the two sides in order to provide a zone of peace or peace-making, in which the opponents could meet and talk, come to know one another, and eventually help one another.

That at least some people from each side did meet and talk, become acquainted, and begin to work together, in the hospitable space provided by the Coalition, proved soon enough that their old quarrel had been not only useless, but also silly, for in answering the question "What do you want?" the two sides found that they wanted mostly the same things: land and water conservation, unspoiled countryside, and biological diversity. They discovered further that in order to have these desirable things they needed each other's help.

The actual work of the Quivira Coalition was defined by another question: was there in fact an irresolvable contradiction between the land economy of the ranchers and the natural health of the land? Years of work and observation on the part of the Coalition, its members, and its consultants strongly suggested the opposite: that there is an indivisible correlation or mutuality between the economic health of the ranch and the ecological health of the land and its waterways. This surprises nobody who has troubled to understand that the human economy as a whole depends entirely upon the wealth and health of the natural world. The ranch's human family and its land must thrive or fail together. The difference is made in the degree of harmony between the human economy and the nature of places. Cattle, for example, can be used invasively, so that they overgraze the grass and trample too heavily the margins of the streams, or they can be managed considerately as components or members of the land's community of creatures.

And so Courtney White's experience with the Quivira Coalition has made him master of two indispensable truths: people of different and apparently opposing interests can work together in goodwill for their mutual good; and, granted their goodwill and good work, a similar reciprocity can be made, in use, between humans and their land.

In the present book, Courtney White, of the American southwest and the Quivira Coalition, recounts his travels, in which he found his work at home confirmed in other places. Not everything from anywhere can be transplanted to somewhere else. But I trust and hope that enough of Courtney's learning in far places will prove to be homework, and we will hear from him again. The Quivira Coalition has made an excellent start on work that is a long way from finished.

PROLOGUE

> "It was the best of times, it was the worst of times, it was the age of wisdom, it was the age of foolishness, it was the epoch of belief, it was the epoch of incredulity, it was the season of Light, it was the season of Darkness, it was the spring of hope, it was the winter of despair, we had everything before us, we had nothing before us, we were all going direct to Heaven, we were all going direct the other way—in short, the period was like the present period . . ."
>
> —CHARLES DICKENS, *A TALE OF TWO CITIES*

THIS BOOK WAS born on a sunny summer day in 2006 when I stepped out of a movie theater with my wife into the warm embrace of a lazy afternoon.

Gen and I had finally found a convenient time to see former vice president Al Gore's inconvenient documentary on global warming, with its dire warnings of environmental and social turmoil ahead if we maintained the status quo. Like millions of others, we were unnerved by what we saw. I was especially disturbed by the graphic images of rising seawater snaking through the streets of Manhattan, Shanghai, and other low-lying cities around the globe. As we stepped off the curb into the parking lot, blinking in the bright sunlight after the movie, I quipped to Gen, "We'd better see Venice, quick."

The film's message wasn't exactly news to us. My work as a conservationist, first with the New Mexico chapter of the Sierra Club and then as a cofounder of the Quivira Coalition, a nonprofit dedicated to building bridges between ranchers, environmentalists, and others around practices that improved land health, had taught me a great deal about the precarious state of our planet. I knew challenges abounded, but Mr. Gore managed to raise my anxiety to a new level. The core issue, I realized, was that sooner or later, Business As Usual would mean serious trouble for every living thing on the planet. Watching the documentary, an image popped into my mind of a bright warning light—in the shape of a thermometer—shining in the dashboard of a speeding vehicle called *Civilization*, accompanied by an insistent and annoying buzzing sound. And like all warning lights, I knew we ignored it at our peril.

As I sat in the dark theater, listening to the former vice president lecture us about our responsibilities and watching the charts and maps of our discontents, I suspected we were seeing only the tip of the iceberg, so to speak. It wasn't just global warming—a great deal more lurked unseen, below the rising waterline. So when Mr. Gore quoted Winston Churchill as describing the run-up to World War II as an "era of consequences"—because Hitler's rise was a pickle of our own making—I immediately thought of the phrase "the age of consequences" to describe our current period.

I mentioned my idea to Gen as we approached the car after the movie. As an archaeologist, I knew she would understand its appeal. History is replete with Eras, Ages, Periods, and Revolutions—Agricultural, Industrial, Technological. Consider all the monikers that have been attached to the current epoch, including the now infamous "Information Age"—infamous because it feels like we're drowning in information while the world unravels. Why not the Age of Consequences? Gen agreed. I filed the thought away.

We climbed into the car and drove home.

The idea to start a chronicle happened a year later, sparked by two events. The first took place over breakfast one morning when our eight-year-old twins, Sterling and Olivia, heard a story on public radio about the possibility of all polar bears dying out as a result of global warming. After a minute or two, the kids froze as they listened, their faces ashen as the disembodied voice of a biologist explained that disappearing sea ice at the North Pole likely spelled doom for the bears. They turned their faces to us, their expressions saying it all: *The polar bears are going to die?*

My heart sank. What could we say? We tried to explain to them that no one really knows if the polar bears are doomed or not. The biologist might be wrong. After all, polar bears have been around for a very long time and have survived a variety of adverse conditions before, including other episodes of severe climate change. Maybe they'll pull through again. This mollified them, and they trooped off to school with their spirits restored.

It didn't mollify me, however.

I turned the incident over in my mind after their departure. What if the polar bears *did* die off? What if Sterling and Olivia never got to see one in the wild, *ever*? Worse, how do you explain to your children what we've done to the planet—to *their* planet—over the past sixty years or so as a consequence of our hard partying? How do you explain to them not only our actions but our inaction as well? It's not enough simply to say that adults behave in complex, confusing, and often contradictory ways, because children today can see the warning light in *Civilization*'s dashboard for themselves. When they point, what do we say?

I didn't know, but finding some way to answer these anguished questions suddenly became a priority.

The second event happened a few months later, while lunching with Wes Jackson and his wife at their home near Salina, Kansas. Wes is founder and director of the Land Institute, which is dedicated to the important business of reinventing the nation's agriculture along regenerative and sustainable lines, so when he said, "We live at the most important moment in human history," I paused between bites of my ham sandwich. That's because a similar thought had occurred to me recently. I asked him what he meant. Wes said that we live at a decisive moment of *action*. The various challenges confronting humanity now require, like a long line of airplanes waiting to land at a busy airport, attention—*immediate* attention. Time is short. Hurry up.

"What sort of action do you recommend?" I asked.

"It means we have to practice restraint," he replied. "That's not something humans do very well, of course. But it's something we've got to learn, or things will get much worse."

Was it possible? I knew that two generations ago, during an era of privation and global conflict, restraint was not only possible but widely practiced. Gas rationing. Victory gardens. Meat twice a week. Prudence and frugality ruled. But everything changed after World War II. The arrow of Progress tipped upward dramatically. We were encouraged at every level to be unrestrained in all that we did—how far we traveled, how much we ate, what we built, or where we sprawled. "Just Do It" became the unofficial motto of my generation, courtesy of a shoe company and an ad agency. Progress, we were told, had no limits and no consequences. Viva la fiesta! Enjoy the party, there won't be a hangover.

They were wrong.

I thought about Sterling and Olivia again. It wasn't anguish I felt this time, however, but indignation. What sort of world will they be inheriting from us? One more bountiful and secure than the one I inherited from my parents, or one more diminished and dangerous? Reports already said that Sterling and Olivia's generation would be less healthy than my generation was at their age—a first in American history, unhappily. Dread began to mix with anger. As a parent, there is perhaps no greater fear than the sense that your children's lives may be worse off than yours. And that's a real worry today, especially knowing it was up to us to handle this important moment in history properly. So far, we weren't doing such a bang-up job.

I know what Wes thinks about it.

Shortly after my visit, I read an essay he wrote based on a commencement address he gave a month earlier at Washington College in Maryland, in which he told the students they were "the children of depletion" and warned them of the inevitable, upcoming contraction of American society. Not surprisingly, the president of the college came rushing up to him after his speech sputtering: "You can't say those things!"

Indeed. That's the trouble with calls to action these days—they can't avoid the umbra of doom and gloom. I've been there myself. In fact, I've heard the mantra of coming trouble so often that I began to suffer from the early signs of what I call "future fatigue." It's a dispiriting affliction that often results in listlessness and apathy. If not caught quickly (usually by sticking one's fingers in one's ears), it can spread quickly, sometimes disabling friends and loved ones. However, when I read Wes's commencement address, I realized that his call to action needed to be heard and shared.

"In painting you this bleak picture, I hope you understand that I am honoring you as adults," he told the students. "You were born on the upslope of energy and economic growth, but much of your life is likely to be on the downslope in the use of nonrenewable energy."

That's because we're depleting the five pools of carbon—soil, wood, coal, oil, and natural gas—at an unsustainable rate, he said. We've burned up, for instance, half the planet's known reserves of oil—one *trillion* barrels—in less than a century. Technology is not likely to ride to the rescue either. Energy, after all, cannot be created or destroyed, just transformed, according to the first law of thermodynamics. So, when sources of energy-rich carbon go into decline, as they will, we either find a suitable replacement, or society goes into decline too.

That's when a second warning light in *Civilization*'s dashboard flickered on in my mind—in the shape of a "low oil" pressure gauge. Urgent action was required here too. Then a third warning light appeared, blinking rapidly. It was the engine warning light, indicating it was time for an overhaul of the main economic means of *Civilization*'s propulsion down the path of Progress.

As a parent and as a writer, the anguish embedded in both of these events created a strong desire to do something beyond my day job with the Quivira Coalition. At the very least, I wanted to document what I was witnessing so that Sterling and Olivia and their cohort could get a sense of why we did what we did—or didn't do—as a society. Hopefully, I would be documenting how we managed to turn off those warning lights in the dashboard. If we failed, however, I was certain that future generations would be asking anguished (and angry) questions of their own. As someone living through this important moment in time, I felt an obligation to chronicle the flow of current events in case it might be useful, now or later. At the same time, I felt compelled to recount my own journey. So, on Earth Day 2008, I began to write, blending headlines, narrative, and observation with travel and research to shape chronological installments, which I posted in an online publication I called *A Chronicle of the Age of Consequences*.

The essays of *concern* in the first half of this book are culled from this project.

Meanwhile, the distress I felt about the world reminded me of a situation a dozen years earlier when I set out to find answers to a different anguished question that had loomed large at the time: was peace possible between ranchers and environmentalists? By the mid-1990s, a bitter feud between ranchers and anti-grazing activists was in full bloom across the American West. Accusing cattle of widespread environmental degradation, activists wanted the cows (and ranchers) gone, pronto, while the ranchers, asserting tradition and legal rights, were equally determined to stay. Their respective hardheaded positions were exemplified by two popular bumper stickers: CATTLE-FREE BY '93! shouted one. CATTLE GALORE BY '94! retorted the other. Lawsuits, countersuits, public denunciations, angry letters to the editor, and general vitriol ruled. It wasn't just the West—the hardheadedness of this particular fight reflected other divides in the nation at the time, including the "red" and "blue" split that would soon engulf our national politics.

I felt a great deal of anguish because I suspected that ranchers and conservationists had more in common than in difference when you stripped away the anger and the politics. Both loved the West's wide-open spaces, wildlife, grass, and water. Both were opposed to the rampant subdivision of private land for housing developments, which had reached a crisis level by the mid-1990s. Nobody won when a ranch was sold and busted up into thirty-five-acre housing tracts, destroying wildlife habitat and a historical way of life simultaneously. While ranchers and environmentalists choked each other to death, real estate developers laughed all the way to the bank.

There had to be another way.

I found it one day when I walked into a statewide meeting of the Sierra Club and saw a cowboy hat sitting on a table. It belonged to Jim Winder, who ranched near Nutt, in southern New Mexico. Jim had joined the Club to find common ground with his purported "enemies," as he put it. Not coincidently, Jim was the only rancher in the state who publicly supported the reintroduction of the endangered Mexican wolf—another white-hot issue of contention at the time. I quickly learned that Jim sincerely sought a resolution to these conflicts. How? It was all about the ecologically friendly style of ranching he used, he told me. Jim bunched his cattle together into one herd and kept them on the move so that any particular patch of ground would be grazed only once a year, mimicking the manner in which bison worked the land. He didn't kill coyotes and didn't mind wolves because bunched-up cows can protect themselves. There was more: because he ranched for rangeland health, Jim got along great with government employees, he had more water in his streams, and, most importantly, he was making money. *Wow.*

Jim's methods worked, as I discovered. It wasn't just Jim either; other ranchers were employing this new style of land management. I decided to label what they did "the New Ranch." I wrote a definition: "The New Ranch describes an emerging progressive ranching movement that operates on the principle that the natural processes that sustain wildlife habitat, biological diversity, and functioning watersheds are the same processes that make land productive for livestock. They are ranches where grasslands are productive and diverse, where erosion has diminished, where streams and springs, once dry, now flow, where wildlife is more abundant, and where landowners are more profitable as a result."

The New Ranch was part of a burgeoning movement in the American West called the "radical center"—a field "beyond ideas of wrongdoing and rightdoing," to quote the poet Rumi. Radical centrist nonprofits and collaborative watershed groups of all sizes grew and spread across the West like wildfire, culminating in a vastly changed political, economic, and ecological landscape in only a few years. I had the honor of engaging with this new movement as a cofounder of the Quivira Coalition. On Spanish colonial maps of the Southwest from the 1600s, "Quivira" designated unexplored territory, which is exactly what we were entering at the time. Following the lead of other "common ground" efforts, we vowed to avoid lawsuits and legislation, sticking instead to the grassroots. It was our belief that the grazing debate needed to start over at the place it mattered most—on the ground.

In early 2003, we gave the radical center a voice when twenty ranchers, conservationists, and scientists gathered in a hotel ballroom in Albuquerque, New Mexico, to craft a document that we originally called a "Declaration of Interdependence" but ultimately titled "An Invitation to Join the Radical Center." It was time for the fighting to end, we declared, and cooperation to begin. As it turned out, the invitation had a substantial impact—the radical center was energized on multiple fronts, and the rancor began to fade as grassroots efforts spread—literally at the level of grass and roots.

Meanwhile, I had begun to energetically explore this new territory, often with my family, and tried to capture what I discovered in a series of essays and profiles, many of which were published in my first book, *Revolution on the Range* (2008). It was an exciting period of time, especially as the radical center took root across the region. The bumper stickers faded, and the vitriol was replaced with a genuine effort to find common ground and move forward. It was very encouraging. In the prologue to *Revolution*, I wrote, "In the years that followed, as the grazing debate faded in the region and as hope and trust began to grow alongside the wildflowers and bunch grasses, an answer to my anguished question began to reveal itself. Ranchers and environmentalists *could* get along, and in places *did*—especially where the dialogue started with soil, grass, and water. Peace, in other words, was possible—and as a result, progress was possible as well."

It's the same with other challenges, including issues associated with climate change. Solutions exist if we're willing to work together and try new ideas (and some old ones). The Quivira Coalition responded to some of these new challenges itself by expanding our work to embrace creek restoration, grass-fed beef production, local food systems, a ranch apprenticeship program, and carbon sequestration in soils. We saw it all as connected—cattle, soil, grass, water, food, people—all working in nature's image of health and regeneration. It was possible, we learned, to balance anguished concerns with hopeful solutions, including many low-tech ones involving sunlight, grass, dirt, creeks, and animals. Too often, however, these solutions are overlooked by members of the media, probably because they are considered too drab to capture the public's imagination. This is unfortunate because they can directly and effectively address the various challenges at work today. Their stories needed to be told, I realized, and if no one else would tell them, I would.

That's why I kept writing.

Eventually, I saw the anguished questions posed by the Age of Consequences and the hopeful answers I found through my work with Quivira to be two sides of the same coin. I decided to pull both into this book. While there's much to worry about these days, I know from experience that there's also a lot that we can do together, beginning at the grassroots—literally, at the grass and the roots. On that point, I'll start with a poem that Wes Jackson used to close his commencement address. It captures this moment in time perfectly—as Wes no doubt had in mind:

"FOR THE CHILDREN,"
BY GARY SNYDER (from his collection *Turtle Island*)

The rising hills, the slopes,
of statistics
lie before us.
The steep climb
of everything, going up,
up, as we all
go down.

In the next century
or the one beyond that,
they say,
are valleys, pastures,
we can meet there in peace
if we make it.

To climb these coming crests
one word to you, to
you and your children:

stay together
learn the flowers
go light

PART ONE

CONCERN

ONE

LATE HIGH FIESTA

(2008)

DURING SPRING BREAK, my family and I drove from Santa Fe to Los Angeles, an old stomping ground for Gen and me, on a classic American road trip, complete with fast food, generic motels, and . . . high gas prices?

Prices had been rising so fast at the pump, in fact, that they gave our sojourn an anachronistic feel, as if we were characters in a diorama in some giant museum. "Look, honey," I could hear someone say, "it's back in that age when people could barely afford to drive their cars long distances!" I mentioned this to Gen as we drove and she laughed, even though she understood that I was only half kidding. She concurs that life in America is in the process of a major historical transition. In fact, the trip illuminated both extremes of the odd, excessive chapter of our national history that I had been calling the Fiesta—the roughly sixty-year period between the conclusion of World War II and the advent of the Age of Consequences (whose specific start date can't be assigned yet).

In archaeology, periods of time that correspond with a particular phase of a civilization's trajectory tend to be labeled Low, Middle, High, or Early, Middle, and Late. For instance, the High Middle Ages took place between 1000 and 1300 AD, and the Late Archaic Period, which in North America was dominated by nomadic bands of hunter-gatherers, lasted from 5,000 to 3,000 years ago. These divisions apply to geology too. The current epoch, called the Holocene, is formally separated into Early, Middle, and Late periods represented by physical changes to the Earth, stratigraphically speaking. In this spirit, I've divided the Fiesta into Low, Middle, and High, with each subdivision lasting approximately twenty years. The crazy boom years since 2000, however, deserve their own category, which I've decided to call Late High Fiesta—as in, party's nearly over! There's nothing scientific to this labeling, just an overwhelming feeling that we live at the apogee of something epochal, transformative, and nearly finished.

At one extreme of Late High Fiesta was Gen's cancer. This was the main purpose of our 2,000-mile expedition, to visit her oncologist and review her progress. Two years earlier, she was diagnosed with a rare form of cancer called carcinoid tumors, a serious but slow-growing disease. It required a specialist and three surgical "interventions" to control, including the removal of a portion of her liver. The result, fortunately, was good news. The latest round of tests revealed her cancer to be much reduced and stable, though she must maintain vigilance (and a monthly shot) for the rest of her life.

Generally, cancer has two sources: a genetic predisposition to the disease or an environmental trigger. In the case of carcinoid tumors, the trigger is something of a medical mystery, though her doctor told us it was likely the consequence of industrial pollution of some sort. In other words, the toxic, synthetic, and increasingly abnormal project we call "civilization" had poisoned Gen. Maybe it was the clouds of DDT that she rode her bike into while growing up in the suburbs of Chicago. Maybe it was the time we spent living in the urban jungle of Los Angeles in the 1980s. Or the power lines near our house after our move to Santa Fe in the early 1990s. Or her proximity to Los Alamos, a nuclear weapons lab, during an archaeological survey she worked on. Whatever the cause, and despite our efforts to live as healthily as possible after our twins were born, the toxic elements of the postwar Fiesta had apparently struck home.

At the other extreme, of course, was Disneyland.

Say what you will about the "Happiest Place on Earth"; it deserves its sobriquet. Constructed in the Low stage of the Fiesta, the park has deliberately oozed optimism and cheerfulness for more than fifty years. It works. Sterling and Olivia love the place, and so do we. In fact, I've been going to Disneyland since I was their age—back in a prehistoric time when you needed a paper "D" or "E" ticket to get on the best rides. Sure, the park is phony, but isn't that the point? We certainly thought so. After a couple of days of confronting cancer, battling mind-boggling traffic, eating food-like substances in generic restaurants, and negotiating a labyrinthine megalopolis (my patience has faded along with my memory of L.A.), we were ready for grinning mice and singing bears.

It worked. We were happy for sixteen straight hours.

I look at it this way: if Late High Fiesta means we have to suffer the consequences of industrial diseases, then why shouldn't we participate in its manufactured happiness as well? After all, a smile is a smile, no matter the source.

You won't hear our children complaining. That's because the organizers of the Fiesta are very good at what they do. On the drive to Los Angeles, for example, we stopped for two days at a sunny resort in Phoenix, Arizona, that featured a wide assortment of watery amusements, including a giant slide, a lazy "river," a noisy waterfall, a relaxing hot tub, and enough chlorine to choke a school of whales. The kids had a blast. They especially enjoyed floating the lazy river atop huge inflated donuts supplied by the resort. Gen relaxed with a book in a poolside chaise while I camped out in the hot tub. What we were doing, of course, was precisely the purpose of the whole Fiesta itself: to relax, indulge yourself, forget about tomorrow. Limits? Consequences? Don't be a killjoy. We'll worry about that later. Meanwhile, party on.

And what a party it's been. Not only have we been going full tilt since World War II, we've come to consider the Fiesta as "normal." I know I do. Take where I live, for example. In the American West, one of the chief intoxicants of the Fiesta was cheap gas, which begat long-distance driving for millions of Westerners. The road had a huge influence on me. It began a few days before my sixth birthday when my parents moved us from Pennsylvania to Phoenix as part of one of the first waves of mass suburban immigration. My earliest memories center on driving—with my father, who worked clear across town; with my friends, who seemed just as restless as I was; and alone, indulging in every minute of my unleaded independence. Cheap, instant mobility became second nature to me.

Historian and author Wallace Stegner once divided the West's denizens into two camps: boomers and stickers. I was not a sticker. I've moved sixteen times since my birth in 1960—a date, by the way, that coincides with the peak of global oil discovery. Look at a graph of the history of oil exploration and discovery, and you will see that it starts in 1859 with Colonel Drake's famous "black gold" gusher in Titusville, Pennsylvania. The rate of new strikes grows steadily from there, with oil fields discovered in Texas, California, and other locations around the nation. Then the graph line rises steeply in the 1930s as the first of the great, giant oil fields are discovered in obscure, far-off nations like Saudi Arabia and Iran. After World War II, major discoveries are commonplace.

In the United States, however, the rate of discovery of new oil had quietly peaked. The first person to notice was a petroleum geologist named M. King Hubbert, who worked for one of the major oil companies. Taking note of this peak in the 1940s, he made a series of mathematical calculations, eventually projecting that U.S. oil production, while still rising at the time, would peak around 1970 and then decline inexorably after that. His calculations proved to be on target. Domestic oil production peaked in 1970 around ten million barrels a day. By 2005, American production had dropped to half that total, including Alaskan oil (also in decline), which is a principle reason why the price of gasoline has risen so sharply, creating a feeling of nostalgia to go along with the anxiety.

It didn't used to be this way, of course.

Gen and I were born toward the end of the Low Fiesta, which began after World War II with the ramp-up of an American economy stoked by cheap fuel, new technology, and access to large quantities of natural resources. Low Fiesta was characterized by an unbounded faith in Progress, the fruits of which became manifest in many levels of American society (think the Clampetts in *The Beverly Hillbillies*). It is symbolized by the roadster, the Interstate Highway System, and the suburb.

We came of age during the Middle Fiesta—a period covering the Yom Kippur War, oil shortages, price spikes, disco, and moral malaise, followed closely by the election of Ronald Reagan to the presidency with his utopian vision of a "shining city on a hill." Forget that self-sacrifice stuff, he insisted, it was morning in America again. He was right—soon morning became a bright, shiny day. We didn't argue. In the late summer of 1980, Gen and I drove from her home in Albuquerque, New Mexico, to Portland, Oregon, for our third year of college. It was a bright, shiny adventure for two young people, freshly acquainted, eager to make our own path together. We traveled in a gas-guzzling yellow Jeep Cherokee, complete with an eight-track stereo. We camped, we hiked, we explored new territory—but mostly, we drove. Gas was cheap and the horizon unlimited, just as it had been all our lives.

By the early 1990s, the Middle Fiesta gave way to the High Fiesta—a full-blown party of awesome scale and noise. It was America's boom time of unlimited growth, Wall Street excess, and Silicon Valley magic. Whatever second thoughts we had as a nation about consequences were replaced by a trance-like condition as the full effect of our oil addiction addled our brains. Questions about limits to growth on a finite planet or climate change, raised by a few brave souls, were drowned out by the roar of partygoers, as the music and the dancing grew louder and more frenzied.

We danced, too.

After college, Gen and I moved to Los Angeles to attend graduate school, hoping to enjoy ourselves and the open air. We drove to the beach, to work, to restaurants, to New Mexico for the summer, to the mountains, anywhere cheap gas and a lead foot would take us. Eventually, we moved to New Mexico for careers, deliberately choosing to leave behind our gridlocked lives in the City of Angels.

The Late High Fiesta coincided with the election of George W. Bush as president of the United States. Not long before, the price of oil had plunged to eleven dollars a barrel, prompting a cover story in *The Economist* magazine titled "Drowning in Oil." The cover photograph showed two workmen, completely drenched in black gold, trying to close a spouting well. This image captured this period perfectly—the party had kicked into high gear and spread like wildfire across the planet. About the same time, Gen and I started a family. We settled into Santa Fe, eventually buying a house. We worked on our careers—Gen as an archaeologist, I as a nonprofit director—before Sterling and Olivia came along. Everything changed after that, of course.

Still, we drove.

As Sterling and Olivia grew, we visited ranches, national parks, beaches, hotels, restaurants, resorts. We drove and drove, and we're still driving. Why not? It's a great time to be alive, frankly. As we departed from the sunny resort in Phoenix, Gen remarked that Western civilization has been aiming at this moment for thousands of years. Just look around us: there's plenty of food and energy, tons of leisure, little heavy labor, few wants, plenty of ease, and lots of toys for young and old alike. Consider what earlier societies had to endure: famine, slavery, oppression, cruelty (as some still do today, alas). Society has worked hard for a long time to create this era of high comfort. It won't give it up without a fight. Neither will many of our cohorts, I bet, despite clear signs of trouble ahead. How our children will feel about all this is an open question.

I await their judgment with trepidation.

There is hope, however. I know because I *heard* it during lunch at Disneyland, of all places. Seeking carbon-based food-like substances for our hungry children, we stopped at a malt shop on Main Street for a fill-up. As we ate, we listened to a talented piano player fill the air with merry ragtime tunes. Soaking up the music, which washed over me like a warm breeze, I was suddenly struck by a thought: while the human species has done a laundry list of rotten things over the centuries and has now put itself and the planet in a perilous spot, we're the same species that invented ragtime.

No other species did it. We did.

We're an inventive and energetic species, full of love and creativity. Watching and listening to the piano player that day, seated in the heart of an iconic temple to the Fiesta, surrounded by friends and family, content, I had another thought: we're up to the challenges ahead.

We have music.

TWO

INDEPENDENCE DAY

(2008)

I HAVE LITTLE doubt that when people look back at the events of the early twenty-first century, many will ask themselves, "What were we thinking?"

I found a partial answer in an unexpected place: the heart of Amish farming country.

On a Fourth of July weekend when a gallon of gasoline cost $4.10 nationwide and a barrel of oil hovered at $140, both records, I found myself among eight thousand Amish farmers in the middle of Ohio watching a parade of brand-new horse-drawn manure spreaders, combines, and hay balers. The occasion was the Fifteenth Annual Horse Progress Days, an enthusiastic celebration of one of the world's oldest energy sources: animal power. This year, the two-day event took place in Mt. Hope, northeast of Columbus. Whether it was intentional or not, the first day fell on the anniversary of the publication of the Declaration of Independence, which seemed fitting.

Although I went mostly out of curiosity, I wasn't unprepared. My family and I had stopped in Mt. Hope the previous summer for a three-day visit with David Kline, farmer, author, publisher, and Amish minister. What I saw and heard during our stay deeply impressed me. I saw a vibrant agrarian community, living and working on a human scale that was wholly alien to me as a child of the suburbs. I saw draft animals at work, manure on the roads, pretty 120-acre farms, smiling faces, and tons of children. And what I heard, when I asked David for a summation of the Amish experience at the end of our visit, was this: "It's all right to live with less."

I needed to know more, so when I discovered that Horse Progress Days would be coming to Mt. Hope that year, I jumped (on a plane, I know). I had spent my youth around horses—but they were the recreational variety. I knew nothing about draft animals or horse farming, except that after being our nation's main power source for centuries, they had become an anachronism by the time of my birth, replaced in a few short decades by the tractor and the oil pump. I wondered: could animal power be a feasible alternative to our dependence on petroleum?

On the evening of my arrival in Mt. Hope, I witnessed the outline of an answer.

Standing at the railing of my two-story B&B, I watched an Amish family bale and load hay in an adjacent field. The hay had been cut a day or two earlier, to dry, and now needed to be "put up" before the increasingly leaden sky began to drizzle. There was a calm, methodical urgency to the family's work. The apparent patriarch of the family, wearing the standard Amish uniform of straw hat, plain shirt, suspenders, black pants, and a beard, stood in a red hay baler that was so old it looked like it belonged in a local history museum. It sounded old too—its single-stroke engine, whose job was to compress the loose hay into a square bale and bind it with string, sputtered and choked so noisily that I expected it to give up and die at any moment.

But it didn't—which was a proper metaphor for the Amish and their work, I suppose.

The baler kept going, pulled by a team of handsome black draft horses that I later learned were Percherons. Together they spiraled steadily toward the center of the field, the baler excreting—for that's what it looked like—a tidy green bale of hay every thirty seconds or so. Not far behind followed another team of horses, guided by a young Amish man—likely a son or son-in-law—who stood on a flatbed wagon. On the ground were three young women, in plain dresses and white bonnets, who loaded the wagon with the freshly minted bales. The work must have been pleasurable because I heard the sounds of talk and laughter from where I stood. When they filled the wagon, the youngsters drove it to a farm across the (very busy) road, returning a short while later to continue their rounds.

In less than an hour, both teams were done. The field had been completely emptied of hay, looking like a shorn sheep, bewildered and turned back to pasture.

I was bewildered, too. *That didn't look so hard to do*, I thought. But my mood changed to astonishment a short time later when I heard the sound of another engine fire up. This was not the sound of a coughing relic, however; it had the confident hum of serious machinery. Indeed, it belonged to a John Deere combine of some sort (I know as much about farm machinery as I do about draft horses). Within a minute or two, it began sweeping across a neighboring hay field of approximately the same size as the Amish field, chased, almost comically, by a tractor pulling a large bin on wheels. The combine sucked up the loose hay from the ground and then spit it through a long pipe into the careening bin beside it. Idling nearby, with their lights on and engines running, were three more tractors with bins, waiting patiently their turn.

In about half the time it took the Amish family to bale and load their hay, the combine had finished its work. All four bins had been filled and the tractors dutifully dispatched someplace over the horizon with their green cargo. The combine, too, took off down the road for parts unknown.

Suddenly, all was quiet.

What had just happened? Two fields of similar size had just been cleared of hay—one principally by horses, the other by horsepower. I wondered: How many gallons of precious diesel had the ancient, coughing red baler used in comparison to the purring combine and speedy tractors? The difference must have been huge. And where did all that industrially gathered hay go? How many miles down the road would it travel to its ultimate destination? I had no idea—but I knew exactly where the Amish hay went—across the road, to be used, I'm sure, to feed the farm's dairy cows this winter. The contrasting images bounced around in my skull. Prodigious vs. frugal. Growling engines vs. laughing voices. Gone vs. stayed.

I soaked up the silence.

The next morning, I drove to David Kline's farm, a few miles away. The day was cool and foggy to the touch. My quite agreeable chore was to transport David and his wife Elsie to the site of the horse fair. The Amish don't drive cars, of course, but they don't mind lifts, especially if there are boxes of books and magazines involved. In addition to managing their organic dairy farm, the Klines edit and publish a charmingly subversive magazine titled, naturally, *Farming*. It's charming because it has a jaunty, colorful feel, full of poems, good stories, good humor, and testimonials to the agrarian life. It's subversive for precisely the same reasons.

It had been a rough year for David. During the winter, he endured a farm accident that hurt his back, and in the spring, he suffered the loss of a sister. I think he looked forward eagerly to the positive energy and camaraderie of Horse Progress Days.

The celebration took place on the fairgrounds of the Mt. Hope auction barn, which covers a large expanse of hilly land. After dropping off David and Elsie, I parked in the vendor's lot, where the black, horse-drawn buggies outnumbered the cars. Taking a moment to wander among the carriages, whose tethered horses dozed dreamily, I noticed that every buggy had a large orange reflective triangle attached to its backside—and I knew why. On the drive from Columbus, I was startled and then annoyed by the number of reckless motorists on the road, many of whom impatiently rode my bumper. But my annoyance turned to alarm when I entered Amish country. The reckless driving continued despite the narrow rural roads, the hills and bends, and the sudden appearance of horses and buggies. Where were these drivers going in such a rush? Why weren't they more respectful, or careful?

I ditched these thoughts and entered the fairgrounds. My eye instantly caught the many hundreds (eventually thousands) of people milling about purposefully, nearly all of whom wore beards and bonnets. Some clustered around the harness and tack vendors. Some strolled through the small village of horse-drawn equipment manufacturers. Some lined up at the food tents, seeking homemade pies and fresh coffee as a brace against the damp air. Some lingered at the logging demonstration site, while others stopped at the farm-animal petting zoo or took rides in carriages pulled by miniature horses through the mud.

Other senses became involved. The smell of BBQ-laden smoke hung in the still air like an aromatic blanket. The wet ground sucked greedily at my shoes. In a distant tent, I could hear the amplified voice of a horse expert instructing a crowd how to gentle a colt. I heard adult laughter and the keening sound of children and the insistent buzz of a woodcutter's chainsaw and the clop-clop of passing horses—huge horses. The biggest I've ever seen. Teams of black Percherons, tawny Belgians, and bay shires, often four abreast, in full harness, clinked merrily as they made their majestic way down a road. They were giant, peaceful horses whose reins rested in the hands of earnest Amish boys, many of whom could not have been more than twelve or thirteen years old.

The whole scene felt medieval, but in a good way. There was a rightness, a humanness, to what I saw, a rightness to the scale, the relationships, the smells, and the laughter. No wonder smiles abounded.

But this wasn't a party. There was serious business going on. I learned this when I attended the demonstration of the latest manure spreaders. It was a remarkable sight. Maybe a thousand Amish farmers, men and women, stood thoughtfully on either side of a freshly plowed strip of soil, watching a steady parade of horse-drawn manure spreaders do their thing. Manure flew high and low as each spreader ran a gauntlet of discerning eyes.

As I snapped photos, I noticed overhead the silent silvery threads of heavy-duty electrical transmission lines. I also noticed a non-Amish farmer in overalls videotaping the action. At nearly five dollars a gallon for diesel, I had little doubt what motivated him.

But it wasn't all gravitas. One vendor had on display a horse treadmill, on which a big Belgian horse walked almost continuously. A shaft connected the treadmill to an adjacent generator in a mock Amish home, where, I swear, it powered a washing machine, a small gristmill, and an icebox. Talk about alternate energy!

In midafternoon, the sun broke through the remnants of the mist, bathing the fairgrounds in bright, sticky light. It reminded me that I had forgotten to bring a hat. By that time, I was getting pretty good at distinguishing fairgoers by their headgear. Bare heads and ball caps revealed non-Amish, of which there was a decent number by lunchtime. Among the Amish themselves, there seemed to be important, though obscure (to me), patterns and variations among the straw hats and white bonnets. David told me there were even significant variations in the choice of suspenders they wore, though he might have been pulling my leg.

By late afternoon, my eyes, ears, and mind were full. Calling it quits for the day—David and Elsie had caught a ride home—I climbed into my rental car and drove in the direction of a hotel in Berlin, a touristy town not far away (the B&B was booked for the night). I was as worn out from walking all day as I was from absorbing the cavalcade of sights and sounds. Reflexively, I turned on the car's radio.

I immediately regretted it.

It was a presidential election year, and the chatter of conservative pundits suddenly filled the car with talk of "energy independence on this Independence Day." Our addiction to foreign oil was shameful, they opined, though not as shameful as the opposition to increased domestic drilling by Democrats in Congress, including a certain presidential aspirant from Illinois. Shame on them, the pundits scolded. In contrast, the Republican candidate for president, they noted proudly, supported new exploration in ecologically sensitive areas. Finally, they prattled, we would get our cherished independence! Yeah right—in a decade, maybe. Here's an idea: why not leave the oil in the ground for our grandchildren instead?

I switched off the radio.

David Kline's words came back to me as I drove: "It's all right to live with less." I think I understood what he meant, finally. By the evidence I saw that day, not only was it *all right* to live with less, it was *possible*, even practical, which is why the farmer in overalls was there with his video camera, I suspect. But it was more than that. Judging from the positive energy that flowed through the fairgrounds, maybe it was even *beneficial* to live with less stuff. Tell it to the pundits, however, or their political cronies in Washington. More drilling is freedom, they argued; increased dependence is independence. And so on.

I shifted uncomfortably in my seat. Everything in our daily lives shouts that it is *not* all right to live with less. Moreover, no one in a responsible social position, including Democrats, says anything to the contrary. And who could blame them? What politician in their right mind would run on a platform of *less*? At the same time, we're beginning to understand that there's only a finite amount of *more* in our future. Some scientists, in fact, insist that we're already in "overshoot"—meaning, we're beyond the point where the planet's limited resources can sustain the current size of the human population (much less the extra two billion projected to be alive by 2050). For many, it may still feel like the Fiesta, in other words, but in reality, the party has already come to an end—we just don't know it yet. Of course, who wants to hear a message like that?

That's why it's comforting to know that seeds of an agrarian alternative, such as horsepower, are out there, being cultivated by knowing hands.

THREE

THE IMPERATIVE

(2008)

I WAS IN the Hall of Biodiversity when the thought hit me like a bullet.

It happened during a business trip to New York City in mid-November, when I found myself with some time on my hands and thought a visit to the American Museum of Natural History would be nice. Passing the famous statue of Teddy Roosevelt on horseback outside, I paused in the rotunda to read some of the former president's words, set in stone:

> "The nation behaves well if it treats the natural resources as assets which it must turn over to the next generation increased; and not impaired in value."
>
> "Character, in the long run, is the decisive factor in the life of an individual and of nations alike."
>
> "A great democracy must be progressive or it will soon cease to be great or a democracy."

Then this from a president universally lauded as a great conservationist:

"Conservation means development as much as it does protection."

I frowned. Wasn't this a contradiction? Perhaps not a century ago, but it certainly seems like one today. But then, human beings are not widely known for the consistency of our ideas or actions, presidents included (I wondered what the Rough Rider would have made of the twenty-first century). In fact, our capacity to hold contradictory thoughts, even to the point of trying to reconcile them, seems to be one of the hallmarks of our evolution. It explains a great deal of our behavior, I thought, as I headed deeper into the museum.

After wandering through huge halls, including one filled with skeletons of mighty woolly mammoths, ferocious fanged mammals, and terrible lizards—all extinct—I drifted into the Hall of Biodiversity about an hour before closing time. I had saved this room for last on purpose, bracing myself for what I anticipated to be a difficult exhibit (the Biodiversity Hall of Shame, I suspected). I followed a large pod of distracted black-clad teenagers into the space, which was split down its length by a wall. In the interstices between coats, sweaters, and heads bowed over ubiquitous cell phones, I glimpsed images of attractive landscapes and healthy animals—all imperiled, I just knew.

Zooming ahead of the slow-moving teenagers, I completed a U-turn at the end of the wall and entered the other half of the room, anticipating gloomy news. Instead, I was surprised to discover a thoughtful analysis of why biodiversity is in trouble and what is being done to fix the situation. On my right was a rather standard litany of trouble: pollution, overcrowding, habitat destruction, and other human sins. When I twisted my head to the left, however, a hopeful solution resided on the opposite wall. For example, there was an educated discussion about careful stewardship of cattle as an antidote to overgrazing—not something I expected to read in the heart of New York City!

The most informative display, however, awaited me at the end of my short walk. On the right wall was a large map of the world, and above it was an electronic ticker. When I arrived, the ticker read, "5000 BCE" (before the Common Era), and the map displayed a handful of lighted red dots, each one representing a million people. They were clustered in the usual places: Egypt, Mesopotamia, China. As I watched, the ticker began to advance . . . 4500 BCE . . . 2500 BCE . . . 500 BCE . . . 0 CE . . . 500 CE . . . and on up to the present day.

You know what happened to the map. New dots appeared, slowly at first, then more quickly, spreading all over the world. Then around 1850, the map exploded. In a blinding blur, red dots swamped nearly all the empty places as the world's population soared exponentially. One billion people by 1800, more than two billion by 1930, six billion by 2000. It was like watching a popcorn machine go totally berserk—but with deep, disturbing implications.

Suddenly, the explosion stopped. There was a pause. Then the ticker reset itself, and nearly all the red dots vanished. Poof! Just like that—the world was sane again.

After blinking in wonder for a moment, my eyes drifted down to a spot on the wall below the map where I saw these words: "The invention of agriculture has caused the human population to soar from 5 million to 6 billion in just 10,000 years. This growth, along with an increase in resource consumption, underlies the great transformation of the world's ecosystems and today's extinction crisis."

That's when the bullet struck.

I spun on my heel and charged into the adjacent Hall of North American Forests. Dodging knots of people, I kept charging, through the Hall of New York State Environment and into the Grand Gallery, a lovely, airy open space. I didn't hesitate. I saw my target dead ahead: the Hall of Human Origins. I knew exactly where I was going—to the reconstruction of a life-size Neanderthal skeleton I had seen earlier in the afternoon.

I needed to confirm my suspicion.

Like many visitors, judging from the big crowds, I had been drawn to the hominid hall by a variety of curiosities, including the perpetually vexing questions: Who are we? Where did we come from? How did we get here? I found some answers too, thanks to a recent renovation to the exhibit (now called the Spitzer Hall of Human Origins). It enabled museum curators to install information from the latest scientific discoveries, including results from new DNA studies on Neanderthals and our specific branch of the hominid tree: *Homo sapiens*.

And what I learned was hair-raising.

First, I was reminded that *Homo sapiens* are the only surviving members of the hominid family left on the planet. Just two million years ago, there were eleven hominid species, including the large *Australopithecus* group, of "Lucy" fame. This branch went extinct about 1.4 million years ago, just as the *Homo* branch started expanding.

Second, I learned that *Homo sapiens* are genetically distinct from the other *Homo* species, including our cousins *Homo neanderthalensis* (though we share 3-5 percent DNA in common). This was news. When I was in grade school, the prevailing theory, as I recall, said we were descendants of Neanderthals—their direct heirs, in a sense. Not so. Now there is conclusive DNA evidence that we were their contemporaries, both branches descended from a common ancestor called *Homo heidelbergensis*, but separate and distinct genetically. We also had important skills that were underdeveloped in Neanderthals despite their larger-sized brains, including better tool-making abilities, an intricate use of language, sophisticated artistic skills, and the cognitive capacity to visualize, and therefore manipulate, the world symbolically (i.e., we were capable of complex, abstract thought).

None of this was good news for our cousins.

Third, I learned that as a subspecies, we came into existence 250,000 years ago (technically as *Homo sapiens sapiens*), which made us the new kids on the block by a long shot. Researchers know the location of our ancestral home too: the original hominid stomping ground of eastern Africa. They also know we became restless quickly. By 115,000 years ago, *Homo sapiens* had expanded into southern Africa. Then, starting around 100,000 years ago, we expanded out of Africa, possibly in two waves, extending first to the Middle East, then into Asia. This was followed by the colonization of what is today Australia, approximately 50,000 years ago. Eventually, we colonized North America as well, via the Bering Land Bridge, coming across approximately 20,000 years ago.

As for Europe, the paleontological record indicates that *Homo sapiens* appeared around 40,000 years ago, possibly during a lull in the Ice Age. Labeled Cro-Magnon by researchers, these ancestors came into contact with our cousins the Neanderthals in a number of different locations—with big consequences.

That's why I headed back to the Neanderthal display—to reread the text. Here's what it said: after 500,000 years of existence, *Homo neanderthalensis* as a distinct line went extinct roughly 10,000 years after contact with Cro-Magnon peoples. Just like that. Was this a coincidence? Not likely. Although there's some dispute among academics about the exact sequence of events, there is little doubt that we had a hand—possibly a *huge* hand—in the Neanderthal's demise. After all, the same fate befell every other hominid species on Earth. It's not clear whether we killed them outright (with our superior technology and cognitive thinking) or outwitted them over time in competition for food and other resources. Either way, the consequence was the same.

Meet *Homo sapiens*, go extinct.

It wasn't just our fellow humans either—a very great variety of animal species died out after contact with our ever-resourceful, ever-expanding ancestors. In Australia, for example, most of the marsupial species larger than a small dog were hunted to extinction after our arrival. A similar fate awaited the flightless birds of New Zealand centuries later. Same with the megafauna of North America (whose skeletons populated the halls of the floors above my head), including mammoths, mastodons, and saber-toothed tigers. While climate change probably played a role in their fate, indisputably so did hunting by an exotic and aggressive species—us. After millions of years of evolution, the megafauna of North America were extinct within a few thousand years of our arrival.

And as I learned back in the Hall of Biodiversity, extinctions continue to this day. In fact, the rate of extinction is rising rapidly, with over 15,000 plant and animal species now at risk of extermination, largely as a consequence of human activity.

As I stood in front of the reconstructed skeleton of my big-browed cousin, thinking about all of this, I suddenly felt like apologizing. Although he was a composite of six separate individuals, he had the wholeness of a real person, though mute, chinless, and extinct. "Sorry about that," I wanted to say. "We knew not what we were doing."

That wasn't quite true, of course. We knew exactly what we were doing. What *Homo sapiens* do is domination, and by extension, extinction—and we do it really well. This was the bullet that struck me in the Hall of Biodiversity: *it's who we are.*

Every exhibit on every floor of the museum practically shouted the same message: we are a dominating species, destined from the start to bend the world to our will whatever the consequences, even if it means extinction.

It felt like an Imperative or a Prime Directive—an irresistible obligation to rule, command, conquer, and dominate. It was our destiny to become a successful and selfish species—not unlike most others, I suppose, only with an exceptional skill set. I'm not saying things were preordained. I doubt that our ancestors, for example, deliberately plotted our cousins' extirpation—but it happened nonetheless, and to our benefit.

It was the Imperative at work.

I departed the museum shortly before closing time and headed outside into the cold, autumnal light of a late-afternoon day in Manhattan. After a chilly walk through Central Park, curiosity got the better of me, and I headed for the nearest bookstore. Once inside, I found a book titled *The Seven Daughters of Eve* by Dr. Bryan Sykes, a research geneticist at Cambridge University in England. His specialty is mitochondrial DNA, which can be used to trace human genetics back, well, to the origin of the species. I took the book back to my hotel room, where I read this confirmation: the Neanderthals were "completely replaced in Europe, and throughout their range, by the technologically and artistically superior new species *Homo sapiens*, represented in Europe by Cro-Magnons. What happened in Europe, as far as we can tell from the genetics, also happened throughout the world, with *Homo sapiens* becoming first the dominant then the only human species, completely eliminating earlier forms." He describes what happened next: "the enslavement of wild animals and plants for food production was the catalyst that enabled *Homo sapiens* to overrun and dominate the earth."

Whew. I closed the book after a while, worn out from so much education.

I decided to turn on the television, though I did so with trepidation. It had been another bad week for American capitalism. The economy was still reeling from the financial atom bomb that detonated on Wall Street in mid-September. The day after my arrival in New York, the Dow sank 445 points to 7,552—an eleven-and-a-half-year low and 52 percent off the high-water mark, set in October 2007. I located one of the cable business channels and settled in for the latest update.

"There is nowhere to run and hide," said the CEO of an investment firm. "This is one for the ages," said another talking head. "It has broken all the rules."

Suddenly, I felt like I was back in the Hall of Biodiversity again.

A week or so earlier, Steven Pearlstein wrote in a *Washington Post* commentary: "The past twenty years have provided ample evidence that uncontrolled flows of private capital have created massive booms and busts that have overwhelmed the financial system and destabilized the global economy. The booms have misallocated capital, widened the gulf between rich and poor, and eroded the norms of behavior that had contributed to social and political harmony.

"There is no denying that American-style capitalism has been undermined by its own success," he continued. "It rewards manipulation over innovation and speculation over genuine value creation . . . And now, through the marvels of global financial markets, they have spread their toxic culture and products to economies across the globe."

That sounded like the Imperative at work. Dominating. All-conquering. Unstoppable.

Still, extinction is *not* inevitable. After all, over the millennia, human societies created cultural checks to this Prime Directive, including ethics, morals, laws, norms, and customs. They are cultural sideboards to our impulses, created to keep the Imperative from running amok—developments that were also in our self-interest. The very idea of progress depends on order; without self-imposed restrictions, anarchy and chaos would rule.

But these sideboards, it seems to me, have weathered badly during the past sixty years—exemplified by the financial crisis currently engulfing the nation. The ethical-moral dike we built over the centuries to hold back a rising sea of industrial greed has sprung so many holes recently that its collapse seems imminent. It wasn't simply the lack of regulation or oversight on the part of our government that caused all the trouble, as some analysts have suggested, but the near-complete failure of our ethics. The Imperative became unrestrained, with serious consequences, as we are discovering. This time, we weren't merely picking on our big-browed Neanderthal cousins—this time, we turned the Imperative on ourselves.

Is it too late? Are we trying to close the barn door after the Four Horsemen have escaped? I don't think so. After all, humans have been down this road before—many times, in fact, as societies rose and fell and rose again. The difference this time, of course, is the scale of the consequences.

I turned off the television, took the elevator down to the lobby, and went for a walk up Broadway Avenue. *It's more complicated than that*, I thought to myself, leaning my shoulder into a bitter wind. Talk of imperatives, ethics, morals, greed, impulses, and so forth doesn't even begin to touch the complexity of our situation. The Age of Consequences is both the product of a pattern of human behavior *and* a set of circumstances peculiar to the past century or so. Our current dilemma is part intention, part accident, and a lot of something in between.

Weary, I stepped into the warmth and comfort of a small diner. I slipped into a booth and ordered a cup of coffee to warm my desert bones. Above my head, a television glowed. On the screen was our newly elected president, Barack Obama, speaking soundlessly to a group of people. He radiated optimism, I thought as I watched. His youthful smile was broad, his face was energized with determination, and his eyes sparkled with intelligence. His whole demeanor, in fact, shone with a serene self-confidence.

It is an amazing moment in American history, I thought to myself. A few weeks later, almost exactly one hundred years to the day after Teddy Roosevelt stepped down as the twenty-sixth president of the United States, Barack Obama, a forty-seven-year-old African American community organizer, would become the forty-fourth executive of this great nation. This was good. I voted for him enthusiastically, and I did so for many reasons, not the least of which was his age—only eleven months separate our birthdays. We're generational compatriots. But after that day's revelations in the museum, I realized there was another reason to feel optimistic about his election: he could lead us in a new direction. We could do things differently, if we wanted to. We could turn the ship of state. That was our prerogative as well.

Nothing was set in stone, except words.

FOUR

THE PARADE

(2009)

AS WE APPROACHED the end of the presidency of George W. Bush, it was tempting to join the chorus of criticism being piled upon our hapless forty-third president. The consensus among historians, pundits, and analysts of various stripes suggested that the best one could say about the Bush presidency was that it was consequential. The harshest assessment was that Bush will be judged by history as one of the worst American presidents ever. Either way, he seemed destined to leave a lasting legacy in his wake.

My take on Number 43 was somewhat different: as the first president of the Age of Consequences (nee the twenty-first century), Mr. Bush was notable more for what he *did not do* than what he did. Additionally, but no less significantly, his presidency told us a lot about the role of the baby boomers in our modern predicament. First, however, I want to start with a recap of our Number 43's accomplishments, such as they are.

What Bush did principally was start two endless wars in Afghanistan and Iraq with all their collateral damage, including over 4,000 dead American soldiers, 30,000 wounded, and a $3 trillion hole punched through the nation's financial spreadsheet. Whether these wars will be judged to have been worth their high cost—i.e., whether they are ultimately "winnable" or produce democratic governments in two nations notorious for their repression and political instability—only time will tell. Recall that the original justification for the invasion of Iraq—Saddam Hussein's purported stockpile of "weapons of mass destruction"—was quickly exposed to be highly erroneous, casting doubt on President Bush's integrity and legacy, I suspect.

What *is* clear, however, at least to this observer, is that the means by which the president and his administration took us to war in Iraq—the falsehoods, the ineptitude, the brazen flaunting of world opinion, the snubbing of allies, the utter disregard for bipartisanship, and an outrageous arrogance in general—is an indelible black mark on his record. Then there were the suspensions of civil liberties, the "extreme rendition" of suspected al-Qaeda members, and the torture of prisoners that followed the invasion. Terrorist threat or no terrorist threat, this wasn't how a United States government was supposed to behave.

Bush didn't seem to care what other people thought. During the run-up to the invasion of Iraq in 2003, as the world reverberated with the sounds of antiwar demonstrations, I remember thinking: can a government call itself democratic if it is completely insensitive to the voices of the people? Not a single protest or voice of dissent mattered—the Bush administration turned a deaf ear to the public and plowed ahead with its agenda. Well, an answer to my question began to take shape in the wake of the Hurricane Katrina disaster in 2005, as an inadequate government response angered the nation. Bush's poll numbers took a dive and never recovered. His fall translated into major gains for the Democratic Party in the 2006 midterm elections, and, ultimately, its triumph in the 2008 general election, in which Bush and his party were soundly repudiated. Apparently in a republic, the voices of the people still matter.

Then there was the economy.

Though much has been made of Bush's strong leadership following the terrorist attacks of 9/11, I believe history will judge his poor handling of the economy as more consequential in the long run. As the economic aftershocks of the meltdown on Wall Street in the fall of 2008 continued to spread across the nation, it was clear that Bush and Co. had no one to blame but themselves and their laissez-faire/free market ideology for the devastation wrecking everything from Wall Street to Main Street.

In an article titled "The $10 Trillion Hangover" in *Harper's Magazine*, economists Linda Bilmes and Joseph Stiglitz wrote: "In the eight years since George W. Bush took office, nearly every component of the U.S. economy has deteriorated." Here's their list of economic woe:

- The nation's budgets, trade deficits, and debts have reached record levels;
- Unemployment is way up, household savings are down;
- Four million manufacturing jobs have evaporated;
- The number of Americans without health insurance has risen 19 percent since 2000;
- Consumer debt has doubled;
- One-fifth of Americans are likely to owe more in mortgage debt than their house is worth;
- Interest on the national debt is now the fourth-largest category in the federal budget;
- The cost of family health-care premiums has jumped 87% since 2000;
- The number of families in poverty rose from 6.4 million to 7.6 million;
- Real median household income dropped 1 percent since 2000;
- Corporate profits surged 68 percent;
- The national debt was $5.7 trillion in 2001; now it is $10.6 trillion—and that's before the impact of the financial meltdown hits;
- Servicing this new debt will cost American families $2,000 a year, year after year, forever;
- The national debt is now 70 percent of the GDP, the highest in fifty years;
- The share of public debt owed to foreign nationals has risen from 31 percent in 2001 to 46 percent today.

"The outgoing administration has made a series of unwise economic choices that together will add up to a burdensome legacy," they wrote. They calculated the bill for Bush-era spending to be $10 trillion, a mammoth and unprecedented number. When Bush took office, he inherited a budget surplus of $128 billion. What happened next, according to Bilmes and Stiglitz, was this: the administration pushed through two massive, inequitable tax cuts and increased government spending by 59 percent, surging the deficit.

"Whether we struggle to break our addiction to deficit spending in order to pay off our debts, or wind up inflating them away," they wrote, "the economic mistakes of the George W. Bush White House will cast a long shadow over the next generation of Americans."

Despite this catastrophic "market failure," the president remained typically unrepentant. In an op-ed published in the *Wall Street Journal* on November 15, 2008, titled "The Surest Path Back to Prosperity," Bush wrote: "The long-term solution to today's problems is sustained economic growth. And the surest path to that growth is free markets and free people . . . The record is unmistakable: If you seek economic growth, social justice, and human dignity, the free market is the way to go. It would be a terrible mistake to allow a few months of crisis to undermine 60 years of success."

Viva la fiesta.

Which brings me to what President Bush *did not* accomplish in eight years: no reform of Business as Usual on any of a variety of big challenges. Take climate change. During Bush's presidency, the United Nations' Intergovernmental Panel on Climate Change (IPCC), a global assembly of climatologists and other scientists, issued two major updates on global warming, one in 2001 and another in 2007. The conclusions of both reports were nearly unequivocal: (1) climate change is largely human-caused, and (2) the situation is unprecedented in recent Earth history, requiring that governments take action to mitigate greenhouse gas emissions or risk the future of life on the planet.

The Bush administration, however, did nothing in response to these reports. At first, they dismissed the science, saying it was "vague" and "contradictory." Then, when the consensus grew too large to ignore, Bush and Co. tacked into the headwind with a new message: action on climate would damage the economy. But as pressure built, including poll numbers showing growing support for action of some sort, the administration tacked once more: all right, we get the message, they said, things are heating up—but now it's *too late* to do anything meaningful, so we'll have to adjust somehow to a warmer planet.

In other words, the first president of the Age of Consequences sat on his hands over a crucial eight-year period, during which concentrations of CO_2, a significant greenhouse gas, rose from 345 to 385 parts per million (it was 280 ppm in 1750). Most scientists agree that if CO_2 concentrations reach 450 ppm, it will cause irreversible environmental damage to the earth's life-support systems—and many think we're already in the danger zone. Either way, I suspect we'll come to regret Bush's inaction in the not-so-distant future.

But we can't just blame George W. Bush for our current troubles. His predecessor, William J. Clinton, didn't accomplish very much either along these lines during his eight years as president. In fact, viewed through the prism of the Age of Consequences, I don't see a great deal of difference between the two men, despite their obvious political and ideological disagreements. Tellingly, during the Clinton years, our corporate industrial economy went global, "free" trade was liberalized internationally, planetary health continued its deterioration, and greenhouse gas emissions maintained their steady rise.

True, Clinton didn't embroil us in endless wars or alienate global public opinion with our foreign policy. And true, he managed to create a federal budget surplus by the time he left office in 2001—a noteworthy accomplishment. He also presided over a strong economy, though cheap energy prices had as much to do with this as anything else. So, one could argue that in comparison, Clinton was a "better" president than Mr. Bush and will enjoy a happier judgment by historians.

Perhaps—but there is another factor at work that will affect both of their legacies.

It's called the baby boom—the seventy-six million Americans born between 1946 and 1964. I'll cite a big reason why this generation is critically important to the dawning of the Age of Consequences: they control much of America's power, wealth, and culture today. This is especially true of those at the front end of the boomer generation, represented by Bush and Clinton, both of whom turned sixty in the summer of 2006.

Which brings me to another reason: I believe there is a significant difference between the front end of the baby boom—those born between 1946 and 1955—and those born on the back end of the bell-shaped curve. And the difference between these two half-generations accounts for a great deal of cultural, social, and economic dynamism (some might say tension) in America today.

I picture the entire baby boom generation as a miles-long parade. The front end of the parade is dominated by a noisy, quarrelsome, revolutionary lot, still determined to change the world. Those of us at the back of the parade—I turned eight in 1968—have a different perspective. We didn't riot in the streets, get gassed, or stage sit-ins. We came of age after Watergate, grew up listening to disco, and surveyed the successes, excesses, and damage created by barricade-crashing front-enders. We picked up the pieces. If front-enders pulled walls down, back-enders tried to build new structures from the rubble.

It is a cliché now to point out that boomers grew up during an era of unprecedented prosperity and security, enjoying the fruits of a robust economy and its unfettered materialism. It's equally cliché to note that many early boomers rebelled against this rising materialism, pushing back against their parents' values with antiwar protests, hippie lifestyles, in-your-face antics, and, of course, sex, drugs, and rock 'n' roll. Revolution was in the air, and tear gas filled the streets. Overturning the old order was the zeitgeist of the day.

These clichés, however, obscure the impressive and varied accomplishments of this determined and idealistic half-generation. They include: the expansion of civil rights; the exercise of democratic rights to freedom of expression and assembly, often under trying circumstances; the challenge to stifling orthodoxies and institutions; the upending of traditional political alliances and orders; the flourishing of the women's liberation and environmental movements; the questioning of materialism; the opening of new frontiers in art and technology; and, of course, a ton of great rock 'n' roll music.

In my opinion, these admirable successes happened for a reason: they were the product of a specific set of front-ender qualities, exemplified by Bush and Clinton, including supreme self-confidence, unstoppable drive, a penchant for confrontation, incredible energy, situational ethics (more pronounced in Clinton's case), and a desire to fundamentally remake the world.

Here's one example: in a chronicle of the era titled *Boom!: Talking about the Sixties: What Happened, How It Shaped Today, Lessons for Tomorrow*, journalist Tom Brokaw tells the story of Dr. Judith Rodin, who became the first female president of an Ivy-League school—her alma mater, the University of Pennsylvania.

Rodin entered Penn as an undergraduate in 1962 and described her first two years as essentially an extension of the 1950s. "Students were very concerned about their own lives, social events, and classroom performance," she told Brokaw. "They were neither politically nor socially active." Then came Kennedy's assassination, the Civil Rights Act, and the expansion of the Vietnam War—and everything changed.

Rodin soon became president of a campus group called Women's Student Government, she organized voter registration drives in the

South during spring break, and she participated in antiwar protests on campus. Upon graduation, she went to Columbia to pursue a PhD in psychology, discovering that some of the older faculty members refused to work with female doctoral candidates because they weren't "serious" about their careers. Then she lost a year's worth of research when a student uprising in 1968 closed the campus and culminated in a violent confrontation with New York City police. The loss of her data made her more focused than ever. "It made me tough in a positive way," she recalled. "It made me very, very determined."

She joined the faculty at Yale and in 1992 became the university's provost. She was, by her own calculation, driven. Her marriage came apart. "I used to think you could have it all," she said. "Now I believe you can have it all, but not all at the same time. There are costs to every decision. Mine weren't cost-free. I had only one child and two divorces. That's a cost."

At a party to celebrate her departure from Yale to become the next president of Penn, she noticed how the young women in the room evaluated her success. "A lot of the women looked at how I led my life and decided they didn't want to live at that level of drive and anxiety, with no free time, forgetting to breathe!" she told Brokaw.

Still, she had no regrets. "When Lyndon Johnson decided not to run for reelection in 1968," she said, "we felt we had changed the world. Whether it was true or not . . . it was an extraordinarily heady experience for my generation, and it influenced us for a long time."

Some chose a different path. Filmmaker Lawrence Kasdan (*The Big Chill*) told Brokaw that at some point, many of his fellow boomers gave up their idealism and succumbed to the impulses of getting rather than giving. "I want the house, I want the car, I want the school," Kasdan said. "No matter how strong your beliefs are, you'll be under constant pressure. It's like trying to hold back the ocean."

When Tom Brokaw reminded Bill Clinton that many of the young people who disdained material goods and wealth in the 1960s were now among the wealthiest people in the world, Clinton replied, "It was always an energetic generation." The real test, the former president said, will be how the generation responds to its wealth.

For her part, Hillary Clinton, who gave the valedictory address at Wellesley in 1969, summed up the message of the 1960s to Brokaw this way: "Choose your own life, make your own decisions. I think it was great for America."

At the end of his book, Brokaw—who is not a boomer—summed up the generation affectionately this way: they were rude, selfish, self-centered, idealistic, and revolutionary. "But it's a different world today," he continued. "We need to be mindful that this is a smaller planet with many more people than forty years ago, and that the future will depend a great deal more on cooperation, large and small, than confrontation."

Which brings me to the back-enders—those born between 1955 and 1964. I believe they developed a very different set of personality traits, and they did so partly as a reaction to front-ender behavior. These traits include: a desire for reconciliation, a need to find pragmatic solutions to problems, a tempered idealism (verging on cynicism at times), and a faith in cooperation. These are qualities, not coincidently, represented by president-elect Barack Obama, who was born in 1961.

It's not just my opinion. In a story published ten days before the inauguration titled "Obama Ushers in a New Cultural Era," reporter Jocelyn Noveck of the *Associated Press* noted that the departure of George W. Bush would mark the passing of an entire generation shaped by the bitter divisions caused by the Vietnam War, civil rights, sexual freedoms, and much more.

"Those experiences, the theory goes, led Boomers, born between 1946 and 1964, to become deeply motivated by ideology and mired in decades-old conflicts," she wrote. And Obama? "He's an example of a new pragmatism: Idealistic but realistic, post-partisan, unthreatened by dissent, eager and able to come up with new ways to solve problems."

Obama himself has commented on this difference. Writing in his autobiography, *The Audacity of Hope*, about the political clashes during the mid-1990s between Bill Clinton and Newt Gingrich, speaker of the U.S. House of Representatives (and another front-ender), Obama said, "I sometimes felt as if I were watching the psychodrama of the baby boom generation—a tale rooted in old grudges and revenge plots hatched on a handful of campuses long ago—played out on the national stage."

He went on to say:

> It's precisely the pursuit of ideological purity, the rigid orthodoxy and the sheer predictability of our current political debate, that keeps us from finding new ways to meet the challenges we face as a country. It's what keeps us locked in "either/or" thinking . . .

> They are out there, I think to myself, those ordinary citizens who have grown up in the midst of all the political and cultural battles, but who have found a way—in their own lives, at least—to make peace with their neighbors, and themselves.
>
> I imagine they are waiting for a politics with the maturity to balance idealism and realism, to distinguish between what can and cannot be compromised, to admit the possibility that the other side might sometimes have a point.

As a back-ender myself, these words resonate very strongly, especially with my experience in the "grazing wars" of the mid-1990s in the American Southwest. In the acrimonious struggle between environmentalists and ranchers over livestock grazing on public lands, I saw a pattern: much of the conflict on both sides was being driven by leaders a half-generation older than myself. They were activists who had come of age during a time when a take-no-prisoners attitude was the norm. On the environmental side, this attitude meant leaders found themselves in unending conflict with rural people. Where I live, in northern New Mexico, environmental activists tangled frequently with low-income, traditional Hispanic communities, especially over logging in national forests. Not surprisingly, more than once, environmentalists were hung in effigy by angry village residents. A bomb was even placed inside the mailbox of a prominent, and highly litigious, environmental group in Santa Fe. Fortunately, it didn't go off. It wasn't any better on the ranching side. A pro-extractive antigovernment group called "People for the West!" was just as aggressive, stubborn, and arrogant.

I don't tell this story in order to assume an air of moral superiority for back-enders. Rather, I think it illustrates how we came to this moment in time. Front-enders, such as Clinton and Bush, had an important role, pro and con, in creating the challenges we all face today. Back-enders will have a big role to play as well, but in a significantly different way. That's because we will eventually assume the reins of power and influence and hold them well into the Age of Consequences. One of us, Barack Obama, already has. And the dynamic tension between these two half-generations will continue to shape a great deal of current history, I believe.

Our current predicament has many sources and is not simply the inevitable result of a human imperative to dominate the planet, abetted by technology and fueled by cheap oil. It is also the product of specific historical forces, including the mighty impact of the baby boom generation. Front-enders, such as George W. Bush, will leave a lasting legacy that has as much to do with his age as it does with his politics. The same will probably be said of Barack Obama. Together, they create important bookends to a momentous period of rapid change in American history.

FIVE

TERRA MADRE

(2009)

LIFE TAKES CURIOUS twists.

I'm a former archaeologist and Sierra Club activist who in 2006 became a dues-paying member of the New Mexico Cattlemen's Association as a producer of local grass-fed beef. Then in 2008, I was selected as an American delegate to Terra Madre, the biennial convening of the "slow food" movement in Turin, Italy, where I joined thousands of farmers from around the planet in a four-day festival of lectures, workshops, and a parade of unbelievably good food.

For a boy raised in the suburbs of Phoenix, Arizona, during the heyday of sprawl, fast food, and disco music, this was a bewildering path. Like everyone else coming of age in a big American city during the 1970s, I didn't give a second thought to anything related to what I ate. Back then, fast food was still considered a *good thing*. Even after I joined the Sierra Club, eventually becoming an activist, I rarely thought about the sources of my daily meals. If I thought about food, it was only in the context of the bad things it did to the land, such as overgrazing by cattle. But there I was at Terra Madre last fall, standing in the lunch line with Peruvian beekeepers, Russian herb farmers, African gourd growers, Italian gastronomists, Scottish students, Indian seed savers, American cooks, Mexican activists, and Chinese academics. Above my head in the cavernous hall—a former

Winter Olympics venue—I could hear the steady beat of global music. On either side of me was a buzz of conversation in the singsong of many languages.

But most amazing of all, everyone was *happy*.

Now, I'm generally an upbeat guy, but this was an infrequent sensation for me professionally. That's because as the director of a nonprofit conservation organization, I get a daily dose of sober headlines: global warming, rising energy costs, population pressures, food riots, wars, the biodiversity crisis, and most recently the financial meltdown on Wall Street (bad news for all nonprofits). Crisis management, it seems, has become part of my job description. That can make for long days and long faces.

That's why Terra Madre was such a pleasant surprise. Smiles were *everywhere*. At one point, I stood in the middle of the giant hall and turned circles in silence; every person I saw radiated positive energy. Many had journeyed thousands of miles to get there, at their own expense, often tracing a personal odyssey. But hardship meant nothing. They were all smiles. The reason, I realized, was simple: they were here to celebrate *food*.

Food binds us together. It is who we are. What we eat, where our food comes from, how it's produced, who grows it, and when it arrives on our table tell us pretty much everything we need to know about ourselves. Our culture is the sum of its edible parts. How we treat the animals that we eat, for example, tells us—or ought to, anyway—a great deal about the state of our nation. Overgrazed range is a food issue. Population is a food issue. Food ties urban to rural, eater to grower, people to land, past to future, one nation to another, our children to ourselves. There is no such thing as a "post-agricultural" society, as author Wendell Berry has noted. We're all eaters. We're all in this together.

It was not a surprise to learn that the slow food movement originated in Italy, where good food is as much a part of the culture as, well, fast food is in America. The two, of course, are connected. Slow food was founded by activist Carlo Petrini in the small town of Bra in 1986 as a deliberate push back against the infiltration of fast food chain restaurants into Italy. His initial aim was to support and defend good food, good eating, and a slow pace of life. The quality of food, Petrini insisted, was intimately linked to the quality of life. "By training our senses to understand and appreciate the pleasure of food," he wrote in a document given to delegates, "we also open our eyes to the world."

Over time, the slow food movement broadened its goals, I learned, arguing that diverse, healthy food is the foundation to overall human well-being and, as a consequence, the very survival of our imperiled planet. Slow food's official mission is to protect, conserve, and defend traditional and sustainable foods, primary ingredients, methods of cultivation and processing, and the biodiversity of cultivated and wild food varieties. This mission is premised on the wisdom of local communities working in harmony with the ecosystems that surround them. Slow food also protects places of historic, artistic, or social value that form a part of our global food heritage.

In 2008, Terra Madre's emphasis was on youth—1,300 young farmers and students attended from ninety-seven countries. The event included the launch of the International Youth Network. The opening ceremony featured an Olympics-style parade of nations, with each delegation dressed in traditional outfits and carrying a placard announcing their homeland. Plenary speakers included Sam Levin, a fifteen-year-old student from Vermont who described his successful effort to start an organic garden on the grounds of his high school. His (youthful) declaration captured the mood of the gathering: "It's a promise to all of you that we will finish what you started," he said. "It's a message to our parents that we will be the generation that will reunite mankind with the earth."

For the next three days, Gen and I wandered in and out of workshops, listened to lectures courtesy of professional translators, browsed the global goods, and braved the jam-packed *Salon de Gusto* in an adjacent building, which featured local food from every part of Italy. We also attended media events, such as the official unveiling of the *Manifesto on Climate Change and the Future of Food Security*, which was self-described as an agroecological response to the challenge posed by climate change. Rising temperatures meant declining food harvests, I read, which in a world already straining to feed itself posed any number of challenges, not the least of which was justice. The poor nations of the world will bear the brunt of climate change's early effects, said the manifesto, which means the rich, polluting nations should bear the brunt of the costs associated with adapting to those effects. It was only fair.

Perhaps most impressive of all, besides the wide diversity of people (and the smiles), was the quantity of youth in attendance. This was a bit of a revelation to me. Youth, I've been told over and over, don't want to go into agriculture anymore. It's not profitable, it's too hard, the hours are too long, and so forth. They wanted jobs in town instead, preferably involving computers. While I harbored doubts about this claim for years, it wasn't until Terra Madre that I realized that an opposite case could be made.

For example, I spoke to one young man who had recently graduated from a university in Montana with a degree in environmental studies. His plan? To become an organic farmer. He had found a farm and was ready to get to work. "Why farming?" I asked him. It was his way, he replied, of doing something about the global challenges confronting us.

"Doing something" was the watchword of Terra Madre. In fact, it could be the motto for the "new agrarianism"—the name being given to this diverse effort taking place around the planet to create an economic alternative to industrialism. Frankly, I find this movement very hopeful and exciting. I left Terra Madre fired up.

▪

BY THE WAY, Gen and I made it to Venice.

We went there first, before the slow food celebration in Turin. It was a lovely fall day when we arrived at Marco Polo Airport—bright, warm, and with a vapory softness to the air that announced, "You're not in arid New Mexico anymore!" We took a crowded water bus to St. Mark's Square (Piazza San Marco) where we disembarked and began to pick our way through the crowds and narrow streets toward our hotel, becoming lost almost immediately.

The first day, we walked from our little hotel near St. Mark's to the Rialto Bridge and back, pausing every hundred yards to gape and gawk. When a slot canyon–like street would empty into a piazza, we would invariably stop for a photo. The buildings were two or three stories tall, often of contrasting styles, and utterly lived-in looking. Every piazza had its own personality too—painted doors, clothes drying on taught lines, peeled masonry, tiny *ristorantes*, impressively serene churches, and tantalizingly shadowed side streets.

Everything was the exact opposite of what I experienced growing up in Phoenix. "Normal" to me was a single-family dwelling sited among a grid of similar cookie-cutter houses, stretching to the horizon. There were no plazas where I lived, no community buildings—unless one considered the mall a communal space—no "heart" to the neighborhood. There was no reason to walk anywhere in Phoenix. Besides, wasn't that why God invented the *car*?

Venetians occupied space very differently than Phoenicians, I saw immediately. Much of this difference is dimensional—Venice is scaled to humans, of course, not cars. That made it attractive on many levels, though some parts bothered me. I missed the sky, for instance, and trees. No matter where one goes in the American West, you are in contact with space, sky, and distance. There are usually plants too, including the ubiquitous Bermuda grass lawn. But in Venice, nature is largely confined to window boxes and an occasional tree. Nature was only a boat ride away, I suppose, but in the heart of the city, it was disturbingly absent.

Venetians ate differently too. I won't go into a litany of contrasts between Italian and American diets, except to say this: I often had no idea what I was eating in Venice. I knew the elements, of course—pasta of various shapes, tomato sauces, cheese, herbs. But the delicious whole was a novelty. Growing up in Phoenix in the 1970s, most of my meals came from just two sources: fast food chain restaurants or the freezer. My mother—a reluctant cook (to put it mildly)—would typically create a meal by pulling meat from the freezer, followed by a frozen vegetable in a plastic bag. The meat would go into the oven, and the plastic bag would go into a pot of boiling water. When supper was finally assembled, we often consumed it in silence in front of the television.

We observed another difference: how Venetians shopped for food. On our second day, we visited the Rialto market. Occupying the same cramped, noisy location it has since medieval times, the market pulsed with energy and vitality. From the smelly fish and gorgeous vegetables to the jostling and constant haggling of the shoppers, it *felt* medieval, at least to this suburban boy. Which meant it felt right—the right human scale, the rightness of fresh food, the rightness of the haggling, even the way the market fit architecturally into its small space. It was as far from the antiseptic supermarket of my youth as one could possibly get. As I said, I thought Phoenix was *normal*.

Not that everything was all sunlight and roses in Venice. At the end of our third day of wandering, Gen and I returned to St. Mark's Square to tour the Doge's Palace, the centuries-old seat of Venetian government. It didn't take long to learn the other side of the story. Venice had a long history of repressive government despite its status as a "republic"—whatever that word meant in the fifteenth century. The worst of the lot was the notorious Council of Ten, who controlled the city's secret police and regularly abused what we would call today the civil rights of the citizenry. Equally chilling were the dungeons, where political dissidents and others were held without trial for long periods of time. In fact, Venice's iconic Bridge of Sighs is so named because prisoners who crossed over it, on their way to the dungeons, would supposedly sigh at their final glimpse of blue sky.

Say what you will about the American suburb; at least it doesn't have dungeons.

After the tour of the palace, Gen and I decided to take a break. We sat down at a small table near the lagoon's edge and ordered two cups of cappuccino. It was a glorious day, warm, sparkling, and timeless. Other than the mammoth cruise liner packed with people that momentarily blocked our view as it glided stately by, it could have been 1750. Venice has been in business for 1,500 years, surviving all sorts of travails—war, oppression, occupation, disease, tourists. It flourished too for quite a while, before entering an inevitable decline (Venice is still shedding residents). It is the epitome of a resilient city, in other words, which is more than one can say about any number of other cities around the globe that had their roots in the ninth century. That's rather impressive, we thought, and probably instructive. Why did Venice survive when other cities collapsed? What makes a city resilient? How long can Venice maintain its resilience in the Age of Consequences?

A short distance from our chairs, the salty lagoon lapped gently against the quay. Much trouble lurks in those gentle waves, I thought. For many years, the main concern of Venetians was that their beloved city might sink into the sea. Venice was built atop millions of logs driven into the mud of a shallow lagoon centuries ago and has been settling slowly ever since. The threat of actually sinking into the lagoon, as a result of this novel architectural strategy, became part of Venice's existence from the get-go. So did flooding (one month after our visit, Venice was hit by its fourth-worst flood in over a century). But the worry these days is of the sea *rising*, not Venice sinking. If the predictions of climatologists are even partially correct, rising sea levels resulting from melting glaciers will claim Venice as one of the first victims of global warming.

Although the threat was tangible to us, sitting there in chairs a few feet from the water's edge, it was hard to imagine this fate on such a lovely day. Everything was perfect. The steady ebb and flow of people on the esplanade behind us, which appeared to include as many Italians as foreigners, was as timeless and unconcerned as the sea. Bells in

St. Mark's Square behind us rang out as they always had, and the sun shone as contentedly on this day as it would a century from now. The wheel of time added another quiet *click* to its endless progression. We sipped our cappuccinos and surveyed our world in silence.

All was well.

That's humanity's conundrum, of course. Our brains tell us one thing—that we're doing insane things to the planet—while a peaceful breeze caresses our faces. I gave a silent mental nod of thanks to my parents' generation for this precise moment: for our comfort, security, freedom, and excellent coffee. Humanity worked hard to get here, doing many wonderful and terrible things along the way, all on display in Venice. And here we are, enjoying the hard-earned fruits of our forbearers' labors. We won't let go of them without a fight.

Does that mean Venice's fate is sealed? Probably. Although Italian authorities are hard at work on various strategies to protect the fabled city from drowning, including the construction of a costly and complicated mechanical barrier at the main entrance to the lagoon, if we don't take significant action on global warming, sea levels are projected to rise three to six feet by the end of this century. Right to where we were sitting. I tried to imagine the legs of our table and chairs disappearing into a few inches of warm water—and decided I didn't want to think about that just then. The day was too beautiful.

I took another sip of cappuccino.

What I can say with confidence is that reconciling what we *know* with what we *feel* may well be the greatest challenge we face in the Age of Consequences.

SIX

WESTWARD HO

(2009)

THIS IS A personal story about manifest destiny.

In 1966, my family and I emigrated from Philadelphia to Phoenix in a covered station wagon, becoming part of a great flood of latter-day pioneers who would change this great nation in ways no one could have imagined at the time. We crossed the Great Plains in a steady caravan of moving vans, sedans, and station wagons—dad behind the wheel, mom navigating, quarrelsome kids in the middle seat, dogs in the back.

We had one goal in mind—opportunity. There were innumerable reasons for leaving home: dank cities, dead-end jobs, misty woods, milk barns, slums, high-rises, boring parents, angry lovers, Eastern snobbery, northern snows, southern humidity, and anything else that humdrummed our lives. Seeking a brighter horizon, we went west as young men and women, drawn by the desert's promise of light, space, warmth, and a swimming pool in every backyard.

We were met with open arms. Homesteading a new land called Suburbia, we were greeted by town leaders who enthusiastically cleared the desert for settlement while their industrious partners planted cheap homes in the newly disturbed soil like row crops. Everywhere we looked, shopping malls and commercial clusters were springing up like patches of flowers (or weeds) after a spring shower. All was fresh, clean, and hopeful.

Clearly, we had found the promised land. Cheap food and gasoline overflowed in conveniently located grocery stores and filling stations; wide, car-friendly boulevards stretched to the edge of the receding wilderness; the dust of a thousand construction projects filled the air like pollination; water flowed magically from our taps despite the near absence of rainfall; seductive carpets of flood-irrigated Bermuda grass lawns tickled our toes; and glorious year-round sunshine fell on our peeling shoulders. Best of all, if it grew too hot while errand-running across the blazing asphalt, we could slip inside our new homes and relax in air-conditioned bliss.

I loved it.

For a young boy, pioneering Suburbia was a great adventure. Our first home backed onto a golf course, and I recall long, restless walks with my mother in late evenings across the trimmed fairways, dodging "tsk-tsking" water cannons and ducking into fairytale forests of oleanders and eucalyptus. A few years later, when we moved across town to a cinder-block house, I discovered the desert. Our new home sat on five acres of backyard wilderness that became both a personal refuge and a stage for elaborate games (alone, alas) that I created among the palo verde trees, creosote bushes, and sandy washes.

Later, we moved again, this time to a townhome in a generic subdivision with no wilderness anywhere. When I went outside to escape various family disharmonies, all I could do was go into the backyard to bounce a ball off the building's sloped roof, over and over, or ride my bike around the cul-de-sacs. The move required that I switch high schools, which disoriented me as much as losing my cherished desert, though it eventually netted me a spot on the soccer team, the presidency of the backpacking club, and a girlfriend.

Soon, we moved again, this time to a spacious house near what was then the last stoplight on the edge of town. I could smell the desert. Liberated at last by a driver's license and a new but mechanically challenged Jeep Cherokee (a source of many adventures in its own right), I began to explore the rapidly expanding boundaries of Suburbia with delight. I dug in archaeological sites with an amateur society, prospected for photographs among the cactus and rattlesnakes, climbed hills, hiked trails, and drove that damn Cherokee back and forth relentlessly on unending blacktopped streets and highways, luxuriating in every unleaded moment.

It was 1976, our nation's bicentennial year, and the world was definitely my oyster.

I never asked, but I'm certain my parents enjoyed their roles as homesteaders too—at least in the beginning. Both had humble roots; my father was born in a shack in a dairy field near Hope, Arkansas, in 1926, and my mother grew up middle class in Charleston, West Virginia. Their journey from want and need to hard-earned success and (for a time) modest affluence was typical of their generation, my father's story especially.

After enduring a hardscrabble childhood spent knocking around Tennessee, North Carolina, and Louisiana with an itinerant dad who at times was a teacher, lumberman, football coach, and preacher, my father determined at a young age to cut a different path. Over his mother's objections, he signed up with the Army, completed a tour of duty in Allied-occupied Berlin, and then attended Vanderbilt University on the GI Bill. Medical school and an MD in neurology followed. After graduating, he won a national award as an up-and-coming doctor, which he parlayed into an opportunity to cofound what is today a highly regarded national center of neurological medicine in Phoenix—a job he held for the rest of his life, earning the accolades of peers and patients alike.

Not bad for a boy born in a dairy field shack.

My mother's journey was no less typical, though it illuminates a darker side of her generation's saga. As a spirited youth, raised in a book-loving but modest and unhappy family (the Great Depression knocked her father back on his heels emotionally as well as financially), my mother yearned to soak up the bright lights of big cities. After marrying my father in 1950, she spent the next decade absorbing every ounce of culture provided by Philadelphia, Los Angeles, Chicago, and other places my father took them to complete his medical training. They attended plays in New York City, vacationed in Boston, traveled to Paris and Prague, all of which made an indelible cosmopolitan impression on her expectations. She especially loved literature and ate up the lives of writers. Judging by the vast quantity and high quality of her correspondence during these years, as I discovered later, I'm certain she harbored ambitions to be a writer herself.

However, things got in her way—children, for instance. My father too, who held old-fashioned opinions about gender roles despite his liberal nature. Then there were my mother's personal demons, including bouts of crippling self-doubt. Part of her situation was beyond her control. As a young woman in the 1950s, she was caught between social riptides, liberation coming ashore and tradition ebbing out to sea. She felt confused, frustrated, and at times angry about both the opportunities and challenges confronting her, as did many women of her generation, I suspect. It also fed her demons.

Phoenix made it all worse. Moving to the suburban frontier in a desert was not on her "to do" list, and after an initial burst of enthusiasm for her new home, she came to resent the city, as well as her fate. Like other pioneering women who "went West" reluctantly but dutifully, leaving the sophisticated "East" far behind, my mother never got over her dislocation or her disappointment. She endured, but not well. She never found the footing she desperately craved in those vigorous times, slipped, and eventually fell.

My father also struggled, especially toward the end of his life, despite his achievements. I think they had trouble keeping pace with the rate of change both in Phoenix and in the world at large. Like many pioneers, my parents were engulfed by the economic fire they helped to light, though I'm certain they didn't see things that way. To my father, it was all progress—which he considered uncritically to be a *good* thing (recall the shack in the dairy field). To my mother, the changes were just part of her general discontent. Progress dog-piled her diminishing expectations, and as a consequence, she recoiled physically and emotionally, eventually embarking on a general retreat. Their home, in fact, became a sort of hermitage from which she emerged only occasionally. By the end of her life, I believe she was content to be engulfed by the city's expanding flames, perhaps hoping to rise again some day from the ashes.

It was much the same with Phoenix itself. What was once a small city with big dreams grew into a big city with big problems—and was ultimately consumed by its own success, though most residents didn't see it that way either, I suspect. Phoenix, too, endured, and not well.

To explain, I want to return to the Old West for a moment. Specifically, I want to review the nineteenth-century idea of manifest destiny and explore its role in the creation of the sixty-year post–World War II economic and cultural blowout of the Fiesta, using Phoenix as a prism.

Manifest destiny was a phrase employed energetically in the mid-nineteenth century by a variety of politicians, journalists, and economic boosters to express the general belief that the United States had an unstoppable destiny to expand from sea to shining sea in accordance with God's manifest will. The term was coined in 1845 by John O'Sullivan, a prominent New York journalist, as part of his argument for the annexation of the Republic of Texas and for American claims to the whole of Oregon, whose northern boundary was disputed by Britain at the time. These claims, he wrote, were logical and necessary "by the right of our manifest destiny to overspread and to possess the whole of the continent which Providence has given us for the development of the great experiment of liberty and federated self-government entrusted to us."

It was a moral call to action that was quickly picked up by less salubrious expansionists who used it to fan the patriotic flames of what became the Mexican-American War in 1846—a conflict that netted California, Nevada, Arizona, New Mexico, and parts of Utah and Colorado for the nation. The clarion call of manifest destiny eventually brought Hawaii and Alaska into the union too, as well as provided cover for our colonial adventures in Cuba, Puerto Rico, and the Philippines at the turn of the twentieth century. It has even been used by some analysts to defend (or criticize) American military adventurism in the twenty-first century, including our wars in Iraq and Afghanistan.

According to historians, one of the reasons why manifest destiny had such a big impact is because it resonated strongly with the concept of American exceptionalism among citizens. This is the idea that America, by virtue of its development as a revolutionary democracy, its novel Constitution, and its perceived divinely directed "destiny" to spread liberty as far and wide as possible, is different from every other nation on the planet, past or present, and thus exempt from the normal rules of history.

The idea that America is exceptional has its roots in the colonial Puritans' vision of a virtuous "shining city on a hill"—a vision that stood in deliberate contrast to the decadence of the recently abandoned Old World. This vision was reinforced by pamphleteer Thomas Paine, who in 1776 argued that the American Revolution was an opportunity, for the first time since the "days of Noah," to "begin the world over again." Abraham Lincoln reiterated this idea in a message to Congress in 1862, arguing that the nation's great experiment in liberty and democracy—the triumph of republicanism over monarchy and oppression—made America "the last, best hope of Earth." In his famous address two years later at Gettysburg, Lincoln would call the Civil War a great test to see if American ideals would survive.

That they did survive that bloody conflagration served to bolster our sense of exceptionalism and destiny, providing a great deal of motivation for much of what Americans did henceforth, including the abolition of slavery and the settling of the American West. These ideals created a desire to extend freedom and democracy not only throughout the continent, but to the world as well, and became, in the process, an important part of our national mission in the twentieth century. Historical events confirmed this calling, from our triumph over Nazi Germany and Imperial Japan in World War II to our victory over the despised Soviet Union in the Cold War and the fall of the Berlin Wall in 1989.

Like a yeasty loaf of bread dough, our sense of exceptionalism kept growing. Mix in our unparalleled economic prosperity, abundant natural resources, a high standard of living, and a huge helping of technological prowess, and you have a recipe for an undisputed American self-confidence that serves millions.

I know, because I saw it all over my hometown.

Phoenix officially came into being on May 4, 1868. The original town site was located on 320 acres of scorching desert. In 1870, the U.S. Census found only 240 people living in what today is called the "Valley of the Sun." By 1950, largely thanks to the invention of air-conditioning, there were over one hundred thousand people within the city limits, plus many more in surrounding communities. There were 148 miles of paved streets. Today, the Phoenix metro area is home to more than four million residents, making it the twelfth-largest city by population in the United States. It covers over five hundred square miles, making it the largest in the nation physically, even beating Los Angeles (at a mere 469 square miles). Since 2000, Phoenix's population has grown by 24 percent, second only to Las Vegas, which grew by nearly 30 percent, and is expected to keep growing by double digits, despite the current economic downturn, well into the future. That sounds like manifest destiny at work to me.

One of my indelible memories of growing up on the edge of Phoenix was the procession of hardware-laded pickup trucks zooming ceaselessly to construction sites everywhere. Festooned with ladders, water igloos, tool boxes, and whatnot, they zipped up and down the fresh streets like bees buzzing around a very large hive. They didn't have to fly far to find nectar either. Cheap housing developments, minimalls, and office complexes exploded across the desert with a fury that had all the hallmarks of an Old West land rush, only without the horses and revolvers. Certainly, the zeal was the same, as was the sense of unstoppable destiny, though perhaps without the religious motivation. Instead, we worshipped a lesser god—Moola—whose divine will directed us to overflow Phoenix with homes, schools, businesses, churches, restaurants, fast-food joints, sports bars, shopping malls, and highways. The only things an Old West miner or cowboy would have missed in 1966 were brothels and livery stables.

If Phoenix in the late 1960s represented a new frontier, marching to the updated tune of manifest destiny, it differed in one important respect from its predecessor: it exhibited a palpable sense of loss. I have a vivid memory from my teenage years of a silent protest. All over the edge of town, numerous real estate signs, each announcing vacant land for sale, had been defaced with a spray-painted lament: save our desert. During a visit one day to a dilapidated horse stable my parents rented way out in the desert, I asked my father what the protest meant. I don't recall his response, but I do recall my feeling of uneasiness, especially as the signs were pushed farther and farther into my beloved desert.

A torn feeling crept into me. I was a suburban kid. I loved all that asphalt and the liberty and convenience it symbolized, especially when behind the wheel of my adventurous Jeep Cherokee. But I also lamented the disappearing desert, its living edge harder to find with each passing month. I understood that my two halves were linked together—one depended on the other—and were like squabbling siblings doomed to quarrel endlessly. As I grew older, however, this torn feeling deepened, until I didn't know what to make of the tension anymore. So I did what many of my peers did to resolve their teenage angst—I moved away and went to college.

The torn feeling nagged at me, however. On trips home, I tried to shield myself from the expanding signs of manifest destiny that I saw everywhere, preferring to cocoon with my parents in their downtown apartment, far from the still-vigorous frontier. It helped that my mother had finally made peace with Phoenix. They now lived close to the main library, the art museum, and other cultural amenities, which had encouraged her to engage once more in the outside world constructively. She became cheerful again, and I recall many happy conversations in their living room revolving around books, authors, movies, and current events.

My father, too, had made peace of a sort with his shortcomings, though not with his deteriorating health. He had contracted adult-onset diabetes in the 1970s, and by the time he was due to retire, his health had declined substantially, requiring daily dialysis treatments. It made him cranky. At the end of their lives, they had reversed roles—my sweet-tempered, generous, optimistic father became grumpy and despondent, while my conflicted, restless, unsatisfied mother mellowed into a cheerful, if still reclusive, angel.

It made for unpredictable visits home.

In a way, their lives continued to reflect the changes consuming Phoenix. Rapid growth, especially the proliferation of new highways in and around the city, created a type of urban-onset diabetes that required daily transfusions of fossil fuel and water to keep the megalopolis alive. It also mocked the proclamation I heard throughout my youth that "We'll never be another Los Angeles!" This type of daily dialysis made residents cranky too, especially those citizens who felt helpless to stop, or even slow, the city's relentless growth. At the same time, Phoenix tried to make peace with itself, or at least with its expectations. It stopped fighting its fate. It stopped pretending it was still a frontier cow town and embraced instead its role as a major cosmopolitan city, with all the traffic congestion and good coffee that came with it. But most of all, it stopped trying to have its desert and eat it too.

It just ate and ate.

It was manifest destiny at work, of course, but it was also the American sense of exceptionalism in action. Not only did we believe in the "rightness" of our cause—to conquer and overspread the continent—we grew increasingly confident that we were exempt from any negative consequences of our actions. If they existed, we were told they either would be (1) fixed by the free market, (2) fixed by government regulation, or (3) pushed far enough into the future to not matter. Phoenix was a perfect illustration. At no time did I hear any second-guessing about limits to growth in a desert. Nothing checked Phoenix's destiny—not concerns about water supplies, cheap gasoline, loss of local agriculture, smog, or what it would take to keep four million people alive in a desert. It was as if we ignored the laws of physics along with the lessons of history.

Progress was good for my parents. They came to a strange land as poor pioneers and prospered along with Phoenix. They lived the American Dream—not the pursuit of material manifestations of success as much as their steady improvement over time. Their lives were better than their parents'; they had more security, more opportunity, more comfort. They didn't do without, go hungry, or stand in unemployment lines; they were well-educated, well-fed, and well-blessed with the fruits of a robust and expanding economy. Best of all, especially for my mother, they could travel, and they saw parts of the globe that deeply impressed them. If they had second thoughts or misgivings about progress, I never heard a word. For them, the future was always bright.

I developed a different perspective. I came of age during the heyday of progress, witnessing the good, the bad, and the ugly. Impressed at first, I have now lived long enough to see that manifest destiny was not necessarily a positive force in our history. I will likely live long enough to see evidence that America is *not* exceptional after all—that despite this nation's many admirable qualities, it is subject to the same historical forces that have worn down all great nations and empires throughout the ages. I know that I've already lived long enough to see us enter the Age of Consequences.

▪

ONE AFTERNOON, WHEN I was thirteen, I rode my mother's favorite horse alone into the desert. Valentine was a huge quarter horse, as sturdy and dependable as she was large. I don't remember why I was allowed to ride alone into the wilderness; perhaps I snuck away. I didn't go far, in any case, just to my favorite place—a remnant of a prehistoric canal that sliced across the desert like a large groove made by a dull knife. It was a subtle feature on the land; to many, I imagined it looked like just another dry wash to cross. But I recognized it as man-made, an artifact of an earlier version of manifest destiny. I loved its subtlety, its muteness, its mysterious origins, and the way it lay on the land. Where did it start? Where did it go? Why was it there?

Our old stable was called Powderhorn, and it sat at the end of a long dirt road in the middle of nowhere, east of town, its only neighbor a funky palm tree plantation. I don't know when the stable had been built, possibly in the 1920s during the heady years of dude ranching and Hollywood Westerns, but by the time we took it over in the mid-1970s, it had lost its shine. Fences sagged, weeds proliferated, old feed rotted in bags, dust gathered, neglect ruled. I loved it. For a thirteen-year-old boy with too much creative energy and not enough friends, the stable became a castle to rule. A fortress to storm. An archaeological ruin to explore. A maze to chase Soviet spies, BB gun in hand. And a backdrop for the nine-hole miniature golf course that I designed and built—whatever it took to keep loneliness at bay.

For my father, then in his midforties, the run-down stable was an important release from various mounting pressures in his life, a diversionary and bottomless bucket of chores—as well as a drain on our meager finances. Unlike many of his generation, however, he was no cowboy wannabe (the horses were for my mother). Still, he loved to tinker, fixing this, building that, earning a tickle of sweat under the desert sun. I suspect the work transported him home to his childhood. Although he had become a neurologist and lived in a big city, the old adage was still true: you can take the boy out of the dairy field, but you can't take the dairy field out of the boy, even in a desert.

Leaving the stable that day, I could tell that Valentine was as eager as I to go exploring. Her huge frame moved along the trail with ease and grace. She spooked at rattlesnakes pretty badly, so I kept an eye out for the coiled menaces. With the other eye, I scanned the horizon. I wasn't playing imaginary games now. No valiant steed or drawn sword in hand. Instead, I opened all my pores and absorbed the desert—the soft wind, the morning smells, the light, the land. I loved the way the desert *looked* empty at first glance, but was actually filled to the brim with life. I had a boyhood fascination with Sonoran cacti that blossomed on these rides. I especially loved saguaros, whose stately, serene forms dot my memory like kindly uncles. I often steered Valentine as close as I could to their barbed, upraised arms, admiring their green skin, which I knew from (careful) experience was smooth and cool to the touch.

Valentine and I dipped into the dull groove of the prehistoric canal, built by the Hohokam people five centuries earlier, riding slowly along its length while I puzzled about the people who had built it. *Why was it so far from any river?* I wondered. *What were they trying to grow here?* Valentine wasn't as curious as I. After an hour, when it became clear that we weren't venturing any deeper into the desert, her thoughts turned toward home, and supper. Yielding to her impatience, I turned her head with one hand and grabbed the pommel of the saddle with the other. I knew what was coming next. As we drew closer to the stable, she picked up speed, despite every attempt I made to prolong our journey. She paced, then trotted, then, because I wasn't careful, galloped. Her bulk and my skinniness meant that I soon abandoned all pretenses that I was in charge anymore. I gripped the pommel with both hands and prayed that I stayed in the saddle until we reached the stable.

These days, the memory of that ride, the open desert, and sturdy Valentine is tinged, not with sadness or regret, but with thankfulness. The stable, the canal, the wilderness are all gone, bladed and buried under a row crop of houses in a tidy subdivision called Powderhorn. But the memory endures. And it's a good memory. I feel fortunate to have experienced rides with Valentine, thankful to see and smell the desert before I became "educated" to the sins of manifest destiny, before the desert disappeared, before progress went on and on. As a youth, I restlessly explored the fine line between city and desert, nature and culture, fascinated by the influence of one on another, the slice of a canal across the land, the contrast of asphalt and desert pavement, a house on a hill, even a golf course set among stately saguaros. It all told a story of expansion and exceptionalism and our disregard for limits. But that knowledge came later. That day, on Valentine's back, all that mattered was the wind.

And freedom.

SEVEN

TO COMPLEXITY AND BEYOND

(2009)

MY SON PLAYS a popular civilization-building computer game that fascinates me. Not only is it exciting—going toe-to-toe with Genghis Khan or Napoleon is never dull—it appeals to the archaeologist in me. Build a civilization from scratch? Cool! But there's another reason I find the game intriguing: it illuminates an important lesson about the Age of Consequences.

You begin the game by selecting a famous empire to command—Babylonian, Greek, Chinese, Roman, Russian, among others. Then you are plopped down in the middle of a vast wilderness circa 4000 BCE and given the mission to build a mighty civilization. You'd better do it quick too, because as many as a dozen not-so-benevolent computer-generated empire-builders soon will be competing against you. To start, you are given a Settler, a Warrior, and, if you're lucky, a Scout—and the race through history is on. As you fend off wild animals and barbarians, your villages grow into hamlets that grow into towns and eventually cities. You gain new technologies over time, beginning with mining, agriculture, hunting, animal husbandry, religion, music, and so forth. Eventually, you discover bronze, iron, math, philosophy, oil, steel, capitalism, environmentalism, and computer technology, becoming in the process a great and enduring civilization.

Of course, the real goal of the game is to wage near-constant warfare. New technologies mean new weapons, and players spend most of their time invading, or repelling, other civilizations. Archers kill enemy spearmen; chariots duke it out with war elephants; knights fight off cavalry; musket men mow down mace men, and are, in turn, bombed by flying dirigibles, and on and on. Meanwhile, you scramble to keep the burgeoning populace happy by building stadiums, libraries, markets, banks, monuments, courthouses, castles, and theaters in your cities as quickly as possible—while praying that you don't run out of gold before the peasantry becomes riotous.

Personally, I like the odd quirks of history that take place during a game. While instructing Queen Victoria to build the Hanging Gardens of Babylon, for example, don't be surprised if a machine gun–toting Mayan chief declares war on you. Sometimes, however, the quirks go too far and give the wrong impression to youngsters. It's disturbing, for instance, when Gandhi declares war on you and invades your territory with his armies, intent on your violent annihilation (seriously, what were the game-makers *thinking*?). Fortunately, Martin Luther King Jr. isn't among the American choices for warlord.

Ultimately, players discover uranium and develop nuclear physics. Soon, they're shooting nuclear missiles at each other while trying to contain the radioactive fallout from missiles shot at them. All civilizations eventually pass into the future, if they survive nuclear holocaust, and a player wins when he or she is the first to land a colonizing party on a planet near the star Alpha Centauri.

Whew.

I've never made it that far, though my son has. I usually bow out around 1800 AD. This is partly due to my consistently poor performance—I never seem to discover gunpowder before Caesar does or monotheism before Mao—but mostly I quit because I'm dismayed by what's happened to the virtual planet I share with my fellow civilizations. The place is a mess. The wilderness has been nuked. The cities are cesspools. All the resources have been exploited. And all I do is fight and fight.

Everything has become bewilderingly complex as well.

This isn't a problem for my son, who has no trouble at all navigating the cascade of new technologies, aggrieved neighbors, sprawling cities, endless trade deals, diverse military units, marauding raiders, sneaky spies, and the occasional natural disaster. I, on the other hand, begin to have trouble distinguishing cannon-wielding Mongolians from grenade-throwing Babylonians from rice-growing Englishmen. As the variables, and the casualties, mount, my brain begins to close its shutters one by one, until it finally sends a signal to my finger to click the "Play Again" button. Presto! A crowded, dreary, and war-weary world is suddenly wilderness again.

And all is quiet and simple.

I bring this up because I think this game sends two important messages to young people: first, progress is benevolent and unstoppable, despite occasional setbacks, and second, decline never happens—technologies never have dark sides and civilizations never collapse, despite what actually happened in history.

On the first point, the game's message is straightforward: Progress is good. History is linear and sequential. Barbarism gives way to sedentism, hunting leads to rocket science, simplicity becomes complexity, and wooden clubs become, well, nuclear bombs. There is a rational sequence to history, suggest the game's designers, and woe to the civilization that doesn't keep up. If you don't develop engineering or chemistry before a rival does, you might be invaded by an army of tank-driving Zulu warriors. And don't forget, the final frontier is space exploration. The race to Alpha Centauri is on!

In other words, the game reinforces our culture's long-standing paradigmatic belief in the inevitability of progress. In the computer game, the arrow of growth and development always points upward. New technology follows new technology, allowing cities to grow, empires to expand, and the rabble to be quieted with sports and art. While the arrow of progress might falter once in a while, due to a shortage of gold, say, or an invading army, sooner or later, progress returns to its relentless upward path. Order is restored. The next new weapon or technology is only a turn or two away. History marches on, we are taught. Corporatism is as inevitable as bronze-making, the alphabet as inevitable as nuclear weapons.

Speaking of which, my son and I have talked a bit about nuclear war. In the abstract, he understands that the nuclear annihilation of a populous city is normally a *bad thing*. But, he asks logically, what is he supposed to do when another civilization begins to fire missiles at him? For my son, nuclear weapons are just another technology, to be discarded when something bigger and badder comes along. We have also talked about what appears to be the game's core message: that history is the sum of new technologies, with a lot of fighting in between. He doesn't see this as a problem, of course. As he views it, the next step up the ladder of progress takes his civilization to a higher, better place. If progress has a downside, it isn't obvious to him, unless it means getting defeated by a nuclear-armed Gandhi.

I have tried to explain to him that in the nonvirtual universe—i.e., the real world—progress isn't so simple or benign (or fun). Progress has consequences, some good, some bad. The invention of the lightbulb was a good thing, but not radioactive fallout. Or slavery. Or colonialism. Or famine. Or climate change. Or global financial meltdowns.

He gets it. However, this conversation doesn't last very long, as you can imagine. No wonder the virtual universe has such appeal to kids these days.

Still, this game isn't helping. While it teaches some useful history lessons and encourages kids to think about the process by which cultures evolve over time, it sidesteps the darker costs of growth and development. This isn't a surprise, of course, because it mirrors a general trend in our society. We adults don't like to contemplate the downside of progress much either—an avoidance reflex aided immensely by industry (including the computer game industry), which likes to distract us from the negative consequences of its actions. Pollution? Biodiversity loss? Resource depletion? Haves vs. have-nots? Come on, forget that stuff. Let's play a game instead—but watch out for Greeks bearing Gatling guns!

Which leads to my second point: there is no such thing as decline in these virtual civilizations. There are no limits to their growth, for example. Natural resources are never exhausted, soil never erodes, climate never changes, corruption never happens, revolutions never take down governments, the social order never breaks up, and so on. Civilizations, in other words, never collapse, despite what actually happened to the Babylonians, Persians, Mayans, Romans, and Mongolians. Empires never really decline either—unless they are beaten in battle. Take the actual British Empire, for instance. Except for those rebellious Americans, it won every major war. But its empire dissolved anyway, causing England to decline precipitously as a world power. I know this makes depressing material for a software game—who, after all, would want to play a game called *Collapse*? Well, some would, I suppose. In the real world, however, it is something to think about.

We need to consider the role of complexity in the rise and fall of civilizations. In my son's game, every civilization becomes increasingly diverse and complicated over time, so much so that I usually bail out soon after the development of "interchangeable parts" or "bureaucracy" or after my medieval knights have all been cut down by a squad of steel-plated tanks. With each turn, the world adds a new layer of complexity—but without the downsides, as I noted. I can't even imagine *that* level of complexity. What would players do if they had to contend with toxic waste, child labor laws, droughts, unemployment lines, lawsuits, epidemics, bad television, banker greed, and so forth? Of course, these conditions are exactly what nonvirtual civilizations have to confront on a regular basis. The light and dark aspects of complexity—lightbulbs and radioactive fallout—with all of their consequences are what can make or break a real civilization.

Including ours.

However, one difference that I see between us and, say, the Romans is the pile of technological choices we have. Thanks to the scientific method—one of the game's stages of development—we have the *possibility* of "high-teching" our way out of serious trouble that previous civilizations lacked. Science also enables us to see the trouble coming in the first place, such as climate change. Whether we act on this foresight by changing our behavior or developing effective high-tech strategies remains to be seen, however. Either way, the future promises to become even more complex, whether we like it or not.

This isn't a problem for my son, or for his friends. A pile of new technologies doesn't bother them at all—in fact, they see it as normal. Perhaps this is a hopeful sign. Increased complexity requires increased skills; if we're going to high-tech our way out of our problems, then we'll need young people versed in the mysterious ways of progress. In this way, I suppose, my son's game is useful. What it teaches him about rising levels of complexity and their consequences (pro and con) can be seen as part of a training program for the future. He'll have a skill set that I can't even imagine, which is a good thing.

However, the real world is not a game. We can't hit the "Play Again" button if events take a nasty turn. We can't save this moment in time and come back to it later if things don't turn out the way we like. We don't get extra lives or free energy sources to use when we want. Instead, we have history lessons that we can choose to heed or ignore as we please. We have human ingenuity at our disposal, to use wisely or not. We have music, art, muscles, compassion, ethics, and faith—in addition to technology—to solve problems and give us hope. But will we use them?

That is the question.

EIGHT

DIRT DAY

(2010)

THE IRONY WAS painful.

On the fortieth anniversary of the original Earth Day, a deepwater oil-drilling rig, aflame in the Gulf of Mexico, sank to the bottom of the sea, triggering one of the worst environmental disasters in American history.

Two days earlier, the rig, called Deepwater Horizon and owned by the oil giant British Petroleum (BP), exploded and caught fire. Fifteen oil workers were injured in the blast, which was caused by a sudden rush of methane gas up from the well site. Eleven others were reported missing and are now presumed dead. BP had been drilling an 18,000-foot exploratory well in the Macondo Prospect, an oil-and-gas deposit located 13,000 feet below the sea floor. The cause of the explosion was determined to be the failure of a so-called "blowout preventer" that should have sealed the broken well after the initial rupture. The full extent of damage caused by the spill, ecologically and economically, is still unknown, but the potential for long-term devastation is high.

When early attempts to shut off the flow of crude came up woefully short, one BP executive complained to the media that working with robots at a depth of 5,000 feet below sea level was like "working in outer space."

Welcome to the Age of Consequences, where our unquenchable appetite for oil now requires us to drill in "outer space"—a place where, when things go wrong, they go *tragically* wrong (the BP executive's comment recalled a famous poster for a 1980s science-fiction movie thriller whose tag line was "In space no one can hear you scream"). If that all feels eerily familiar . . . well, it is. Part of the painful irony of the Deepwater Horizon spill is that *another* oil spill played a key role in goosing the original Earth Day into existence.

On January 29, 1969, a Union Oil drilling rig located six miles off the coast of Santa Barbara, California, suffered a similar natural gas "blowout." Unlike BP, however, Union Oil was able to cap the blowout quickly because they were working in shallower water. Unfortunately, high pressure in the oil field created ruptures in the ocean floor, causing 200,000 gallons of crude to leak into the ocean. Currents quickly spread thick, oily tar across miles of pristine southern California beaches (the BP spill, in comparison, leaked 200,000 gallons of oil in only a few hours). After eleven frenetic days, Union Oil stopped the leak, but the damage had been done.

Seabirds were especially hard hit—approximately 3,600 died as a consequence of the spill. Worse for Union Oil, images of dead birds, seals, and dolphins were repeatedly played on television sets across the nation, causing a storm of outrage. People burned Union Oil credit cards, boycotted their gas stations, and gathered 100,000 signatures on a petition that called for a ban on offshore oil drilling. "I am amazed at the publicity for the loss of a few birds," is how Union Oil chief Fred Hartley responded. "Never in my long lifetime have I ever seen such an aroused populace at the grassroots level," said Thomas Storke, editor of a Santa Barbara newspaper.

And this from President Richard Nixon: "The Santa Barbara incident has frankly touched the conscience of the American people."

As if on cue, in June of the same year, a portion of the highly polluted Cuyahoga River in Cleveland, Ohio, burst into flames.

The following spring, Earth Day rocked the nation.

A blizzard of landmark federal legislation followed in the wake of these events, including the National Environmental Policy Act (1970), an extension of the Clean Air Act (1970), the Clean Water Act (1972), the Endangered Species Act (1973), and the creation of the Environmental Protection Agency, all signed into law by President Nixon.

In a speech in 1970, Nixon told the nation, "The great question . . . is, shall we make our peace with nature and begin to make reparations for the damage we have done to our air, our land, and our water? Restoring nature to its natural state is a cause beyond party and beyond factions . . . It is a cause of particular concern to young Americans—*because they more than we will reap the grim consequences of our failure to act on programs which are needed now if we are to prevent disaster later*" (emphasis added).

Nixon continued: "The argument is increasingly heard that a fundamental contradiction has arisen between economic growth and the quality of life, so that to have one we must forsake the other. The answer is not to abandon growth, but to redirect it . . . I propose that before these problems become insoluble, the nation develop a national growth policy. Our purpose will be to find those means by which federal, state, and local government can influence the course of . . . growth so as to positively affect the quality of American life."

These words are amazing words to read today, four decades later, and not simply because they are the words of a *Republican* president. They are amazing because they were so prescient. We didn't act on them and now must bear the consequences, as Nixon warned. As the Deepwater Horizon disaster demonstrates, America never developed a national growth policy, unless you consider unchecked growth a policy. And Deepwater Horizon is just the tip of the iceberg, to mix crisis metaphors.

I came across Nixon's remarkable words in the transcript of a speech given this past January by Dr. Gus Speth, who recently retired as the dean of Yale's famous School of Forestry. A former leader in the environmental movement and aide to President Jimmy Carter, in the past few years, Speth's attention has focused on our converging planetary crises and how the shortcomings of the environmental movement failed to prevent them. In fact, he argues that the movement, catapulted into the front lines of American politics and society during the original Earth Day, is now dying—and doing so precisely at the wrong moment.

"The environmental movement," said Dr. Speth, "has grown in strength and sophistication, and yet the environment continues to go downhill, fast. If we look at real-world conditions and trends, we see that we are winning victories but losing the planet, to the point that a ruined world looms as a real prospect for our children and grandchildren."

He cites evidence from the United States:

- 40 percent of U.S. fish species are threatened with extinction, as are a third of plants and amphibians, and 20 percent of birds and mammals.
- Since the first Earth Day, we have increased the miles of paved roads by 50 percent and tripled the total miles driven.
- Half our lakes and a third of our rivers still fail to meet the fishable and swimmable standards that the Clean Water Act said should be met by 1983.
- A third of Americans live in counties that fail to meet EPA air-quality standards.
- And we are still releasing truly vast quantities of toxic chemicals into the environment—over five billion pounds a year, at least.

And the global news is even grimmer:

- Half the world's tropical and temperate forests are now gone, with the rate of deforestation in the tropics continuing at about an acre a second.
- Almost half of the world's corals are either lost or severely threatened.
- Species are disappearing at rates about one thousand times faster than normal. The planet has not seen such a spasm of extinction in sixty-five million years.
- Over half the agricultural land in drier regions suffers from some degree of deterioration and desertification.
- Despite stern warnings now thirty years old, we have neglected to act to halt the buildup of greenhouse gases in the atmosphere and are now well beyond safe concentrations.
- The following rivers no longer reach oceans in the dry season: the Colorado, Yellow, Ganges, and Nile, among others.

"And so here we are, forty years after the burst of energy and hope at the first Earth Day," Speth said in the speech, "on the brink of ruining the planet. Indeed all we have to do to destroy the planet's climate and biota and leave a ruined world to our children and grandchildren is to keep doing exactly what we are doing today, with no growth in the human population or the world economy."

And he said this *before* BP's oil rig exploded and sank into the Gulf of Mexico.

For a simple reason, he includes the environmental movement in the business-as-usual paradigm that is destroying the planet: it has been ineffective in stopping these developments. What is needed now, he says, is a *new environmentalism*, one that confronts our destructive economy head-on.

A good place to start, I think, is changing the name of Earth Day to "Dirt Day."

The idea came to me after reading *Dirt: The Erosion of Civilizations*, by David R. Montgomery, a professor of geology at the University of Washington. Although ostensibly a history of dirt, it's really a book about the failure of modern society to learn lessons and avoid repeating mistakes that hastened the demise of past civilizations. Most modern crises, Montgomery observes, have an element of speed to them: an oil spill, for example, or an accelerating rate of species extinction, and so on. Dirt is different. Its crisis timescales are slow—almost too slow for humans to see. Erosion, for example, can be going on for decades before it's noticed. And by then, it is often too late.

Dirt is created by the weathering of solid rock over time, usually accumulating at the rate of about one inch per one thousand years. In some agricultural systems (of the sustainable variety), this rate can speed up to one inch per one hundred years, largely by the application of manure and other natural fertilizers that help to create rich, thick humus. Unfortunately, we are eroding our soil in many places around the planet at the alarming rate of an inch per *decade*. Ultimately, this will result in serious trouble.

The reason is simple: there is no substitute for dirt. Oil and natural gas can perhaps be replaced by other energy sources, preferably renewable ones, but plants and animals—and us by extension—require dirt for our existence. Nothing else does what dirt does. In addition to being the medium in which our food grows, dirt cradles drinking water, recycles dead material into new life, circulates essential nutrients, stores carbon, and even remediates waste. Montgomery calls it our most underappreciated and yet essential natural resource. If we wash it away, then we pay the consequences.

As a crisis, however, dirt is too hard to fathom. That's because it never *feels* like a crisis; instead, it just quietly slips away, one particle at a time. In this way, it is similar to the climate crisis, in which the steady drip-drip of greenhouse gas buildup in the atmosphere isn't noticed on a daily or monthly basis until its cumulative effects can be detected—often too late. As Montgomery notes, it is this type of drip-drip crisis that poses our greatest challenge.

"Mortgaging our grandchildren's future by consuming soil faster than it forms," he writes, "we face the dilemma that sometimes the slowest changes prove the most difficult to stop."

Few places on Earth produce soil fast enough to sustain industrial agriculture over human time scales, he argues, which means we are slowly running out of dirt. This isn't a new phenomenon, of course. As the Sumerians, Greeks, Romans, Mayans, Chinese, and early settlers in America could tell you, dirt matters. In fact, the history of dirt suggests that how people treat their soil can impose a life span on civilizations. That's because dirt is required for food production and water retention—activities mandatory for a healthy civilization. Time and again over the course of human history, social and political conflicts grow when there are more people to feed than can be supported by the land.

Civilizations don't disappear overnight, and they don't choose to fail. More often, they falter and then decline as their soil disappears over generations. Rome didn't so much collapse as crumble, says Montgomery, wearing away as erosion sapped the productivity of its homeland.

"A common lesson of the ancient empires of the Old and New Worlds," he writes, "is that even innovative adaptations cannot make up for a lack of fertile soil to sustain increased productivity. As long as people take care of their land, the land can sustain them."

Conversely, neglect of the basic health of the soil accelerated the downfall of civilization after civilization even as the harsh consequences of erosion and soil exhaustion helped push Western society from Mesopotamia to Greece, Rome, and beyond. It's not just ancient history either. In America today, millions of tons of topsoil are eroded annually from farm fields in the Mississippi River Basin into the Gulf of Mexico. America's farms erode enough soil every year to fill a pickup truck for every family in the country.

Worldwide, an estimated twenty-four billion tons of soil are lost annually, several tons for each person on the planet. Over two billion acres of virgin land have been plowed and brought into agricultural use since 1860. Until the last decades of the twentieth century, clearing new land compensated for loss of agricultural land. The United Nations estimates that 38 percent of global cropland has been seriously degraded since World War II. In the agricultural era, nearly a third of the world's potentially farmable land has been lost to erosion, most of it in the past forty years. According to Montgomery, this happens because cultivating a field year after year without effective soil conservation is like running a factory at full tilt without investing in maintenance or repair. Good management can improve agricultural soils just as easily as bad management can destroy them, though once soil is lost, recovery generally lies beyond the horizon. With just a couple of feet of soil standing between prosperity and desolation, civilizations that plow through their soil vanish.

Perhaps just as importantly, technology will not save us. It can't create more dirt; only nature can. This is the big lesson of dirt, Montgomery says: when you depend on a resource that is difficult to renew quickly, eventually you wind up in serious trouble. Modern society fosters the notion that technology will provide solutions to almost any problem, he writes. But no matter how fervently we believe in its power to improve our lives, technology simply cannot solve the problem of consuming a resource faster than we generate it: someday, we will run out of it.

According to Montgomery, even a casual reading of history shows that under the right circumstances, any combination of political turmoil, climate extremes, or resource abuse can bring down a society. Alarmingly, he says, we face the potential convergence of all three in the upcoming century as shifting climate patterns and depleted oil supplies collide with accelerated soil erosion and loss of farmland. Should world fertilizer or food production falter, political stability could hardly endure. We must do things differently.

"Clearly, more of the same won't work," he concludes, sounding very Nixon-like. "Projecting past practices into the future offers a recipe for failure. We need a new agricultural model, a new farming philosophy. We need another agricultural revolution."

We need a Dirt Day.

NINE

LIFE IS GREAT

(2010)

WITH A FLICK of the switch, I banish the darkness.

It's four a.m. on a Monday—time to get some work done before the sun, or the kids, stir. In the bathroom, I twist both faucet handles at the sink and watch groggily for a few seconds as the water twirls merrily down the drain. Where does this water from come? An ancient aquifer nearby, as I recall. Can't be rainwater, I say to myself as I splash water onto my face in an attempt to ward off a desire to go back to bed; we only get twelve inches of precipitation a year here, if we're lucky. Which reminds me. Drying my face with a cotton towel, fresh from yesterday's laundry, I make a mental note to buy rain barrels for our roof's downspouts, adding it to a lengthy to-do list.

Leaving the bathroom, I wend my way into the kitchen, where I make an unsteady beeline for the coffeemaker. I didn't touch a drop of the evil brew until I was thirty-one, giving in only after a move to our home at seven thousand feet and a subsequent snow storm that winter. I grew up in the desert and lived in Los Angeles for years, so snow was a difficult concept for me to grasp initially, requiring what has since become a comfort food—a warm cup of coffee. In any case, I am grateful that a steady and apparently endless supply of the evil roast is available to someone who lives far, far away from a coffee plantation.

If there were a coffee god, my daily ritual would include an oblation of thanksgiving, perhaps in the form of a teaspoon of sugar.

Mug in hand, I drift into the living room and settle into a chair at the computer desk, waiting for the caffeine to work its magic. Although it's not quite summer yet, the windows are cracked open enough to let dryland smells into the house. It's a remarkable privilege to live here in this beautiful place, in what geographers call a high, cold desert. Prehistorically, there was only enough food, water, wood, and arable land to support small populations of people, most of who had to move frequently to find fresh resources or dodge a drought. It's totally different today, of course—except for the same sparse amounts of local food, water, fuel, and arable land. Thankfully we have oil, without which I wouldn't be able to live here. Perhaps another offering is in order, this time to the gods of petrochemicals, who we never, ever want to anger.

I punch my laptop on. The computer is another of life's confounding miracles. I never knew I needed one—and now can't imagine my life without it. When Gen and I were in college in the early 1980s, the personal computer had not yet made its debut on campus. That meant we had to rely on *typewriters* to graduate. This fact makes us absolutely archaic to our children and their tech-obsessed friends. There were no mobile phones either, I tell them a bit proudly, or iPods, though we had something called a Sony Walkman for music. Typewriters? Cassettes? Landlines? The whole idea gives them the shudders. As my laptop finishes its warm-up cycle, this line of thought provokes a question: What did we actually *do* with all that free time way back then? I recall that we skied, and hiked, and drove to the beach. Watched TV. Read books. Cooked. Talked. *Old people* stuff. You know, ancient history.

The computer is ready to go. As I scoot my chair closer to the table, another question flits through my mind: Where does our electricity come from? Not nearby, like our water, that's for certain. Is it from the nuclear power plant near Phoenix? The coal-fired plant near Four Corners? The natural-gas facility near El Paso? The big dam on the Colorado River? A wind farm near Albuquerque? All of the above?

Two hours later, I shut down the computer, rise from the chair, stretch my stiff muscles, then stride purposefully toward the kitchen to start the breakfast marathon. I switch on a lot of lights, even though the dawn is brightening quickly outside. I stab our old radio to life and reel instantly at the news: terror threats, political gridlock, greed, avarice, unemployment, upcoming elections. After a few minutes, I stab the radio off, not wanting to scare the kids. I switch on the CD player instead, filling the kitchen with the reassuring strains of a Mozart concerto. Then I turn to the main event of the morning: the breakfast menu.

Like many of their friends, our twins will only eat from a short list of acceptable items, very few of which correspond with anyone else's preferences, necessitating a kind of daily food ballet. For example, our daughter likes sausage, which we buy organically and locally, soaked in maple syrup from Vermont. She'll eat English muffins too. But our son won't touch either one. He prefers industrially produced corn dogs, which no one else will eat (for various reasons). However, he likes Mexican meals, so burritos are popular in our house—except with our daughter. Gen prefers granola with yoghurt, or polenta, or Irish oatmeal, none of which the kids will touch. I like eggs, which we procure from our small flock of chickens in the backyard. Gen loves them too, but the kids told us the other day that they are tired of eggs. Our daughter still likes homemade waffles, though our son is tired of them as well. He'll eat fried potatoes, but she won't. She likes cereal, but he doesn't, of course. They are united, however, in their opposition to anything green at suppertime, which we force them to eat anyway. We do agree on organic milk, butter, hamburger, pasta, and rice, fortunately. Otherwise, we might starve.

We won't starve, of course. That's because our food system is a miracle, I think to myself as I pull a package of frozen sausage from the freezer and place it in the microwave oven for defrosting. We can eat what we want—or refuse what we want—from wherever we want, at any time we want. Peaches in February? No problem. Shrimp in a high, cold desert? No sweat. Coffee from an obscure island in the South Pacific, chocolate from Europe, lettuce from California, plasticware from China, honey from Albuquerque, canned green beans from God knows where? No problem. Even the microwave is a miracle. Look: the sausage is defrosted in a minute, ready for frying. I pull out a nonstick pan—another miracle—and place it on the stove. Hash browns, eggs, English muffins, marmalade, corn dogs, sliced cheese and meat for lunches, and sandwich bread quickly follow. It makes for a heap of food on the kitchen table, suggesting that a prayer to the food gods is probably in order as well.

But I don't have time to think about divinity this morning. Everyone is late getting out of bed. There's a traffic jam at the shower, and no one is making his or her lunch yet. I set the table, feed the dog, and turn down the Mozart when a minor dispute erupts between the kids over the last remaining berry-flavored juice drink. We flip a coin. "Honestly," I say to myself as the juice drink is dispatched to a lunch box triumphantly, "what would our grandparents have thought of such an argument?" What would they have thought of our breakfast in general? They would have considered it a miracle as well, I'm sure, though they might have been appalled by how much food ends up in the trash can or the chicken bowl. Waste not, want not, as Gen's mother used to say, and pass the marmalade, please.

We gather at the dining room table to eat, finally, though instead of saying grace, we talk about the day's calendar: what's happening in fifth grade today, who has what music or karate lesson where and when after school, what needs to be done on the homework front, and so forth. The discussion is complicated by my imminent departure on a business trip, which the kids have forgotten. Faces fall, but only for a moment. I'll be right back, I tell them. Zipping to and fro across the country makes air travel feel like just another item on the daily chore list, except for the hugs and kisses. But I don't have time to contemplate that right now. I need to clear away the dishes.

Shortly before eight a.m., Gen and the kids trundle out to the car and head off to school and work respectively. We ought to walk; after all, school is only half a mile away on an easy path, but for reasons that are not entirely clear to me, we don't. We drive, like every other family, filling up the school's too-small parking lot with every conceivable model of SUV and minivan known to humanity, often piloted by solitary parents in a hurry (and on cell phones). Sometimes I take a quick peek at the bicycle rack outside the library to see if anyone has ridden to school that day. The district recently expanded the elementary program from K–6 to K–8, and one might logically expect an unusually adventuresome seventh or eighth grader to ride his or her bike to school. Alas, it's always empty. And why not? Gas is cheap, cars are convenient, we're in a rush, and kids don't like to walk much anymore. It makes perfect sense.

After a final round of good-bye kisses, the family pulls out, and I retreat to the kitchen to put things away. Later, after some bill paying, a walk with the dogs, and a shower, I settle down with a stack of maps and guidebooks to Europe. Gen and I turn fifty this fall, and we've decided to treat ourselves and the kids to a whirlwind tour of Rome, Venice, and Paris, with lots of Roman ruins and medieval castles in between. Ever since Gen and I visited Venice, it's been a dream of mine to celebrate my birthday alongside the Rialto Bridge, which I'm determined to fulfill. Why not? Other than the expense, it's easy to get to Europe, and once you're there, it's easy to get around. My plan is to use it all: planes, trains, buses, taxis, boats, and a rental car. Everything is in the guidebooks—where to go, what to eat, where to sleep. Besides, it'll be a history lesson for the kids. Us too—a firsthand look at Western civilization, including centuries of wars, hardships, political upheavals, religious rifts, technological breakthroughs, economic strife, and social progress . . . all so we can watch cable TV, surf the Internet, goof off with video games, and get diabetes and cancer.

And travel to Europe.

I put the map and guidebooks away, pack my travel bag quickly, check on the chickens, and head out the door. I jump into the truck and head into town, where I need to put in time at the day job and run a few errands before catching my flight. It's a beautiful day, bright, clear, and warm. Mid-May at seven thousand feet in a cold desert is full of the promise of summer. I can't wait for the T-shirt weather to begin. I slip onto the interstate and nestle into the flow of traffic that will carry me the short distance to town. I scrupulously obey the speed limit, despite knowing what's coming. Soon, my bucolic attitude evaporates as a steady stream of cars and trucks speed past me. I notice that nearly every vehicle carries only one person (mine included). A few drivers talk on their cell phones, but most zip along in air-conditioned isolation. I'd prefer to take a bus to work, but there's no practical public transportation available, despite high gasoline prices (nearly three dollars a gallon today). That speaks volumes about our priorities as a society, I think.

A few minutes later, I pull into the small graveled parking lot at my office. After a round of "good mornings" to coworkers, I climb the stairs to my office, which still smells faintly of leather. Last fall, in a fit of indulgence, I purchased a nice leather couch to place below a window. Seeing it this morning, I suddenly wonder: Where did it come from, who made it, and what is inside it? Most of my office is populated with things that have a history—an old wooden table inherited from my parents, bookcases from our days in L.A., framed photographs from a previous attempt at a career, maps accumulated from various trips, photographs of the family, and a wide variety of knickknacks picked up over the years from all corners of the American West. They make the mass-produced leather couch an anomaly (or is it the other way around?). All I had to do was pull out the credit card and pay for it. Like so many things in our lives today, its purchase was fast, convenient, and anonymous.

I settle down to work, which means I must stare, once more, deep into a computer screen. Perhaps because I grew up in an archaic age, I stubbornly resist being sucked into the virtual 24/7 world that has consumed so much of our society. I'm still an eight-to-five guy, which means I don't do much email on the weekends and I don't do social media at all (no Twitter or Facebook accounts for me). My cell phone is just a phone. It doesn't entertain me, check the stock market, or cook supper. I haven't even programmed it with the phone numbers of friends and family. I'm required to memorize their numbers. That's all right. I'm trying to inhabit as much of the 3-D universe as possible, fearful that our expanding obsession with the 2-D world is setting us up for a major fall. But that's another topic for another day.

At noon, I shut down everything, pack up, say some quick goodbyes, and jump back into the truck. I need to run a few errands in town, starting with a pit stop at the bank to cash a check. My next stop is a natural foods grocery store. I need snacks for the trip. Cruising briskly down the aisles, I realize the store is another mundane miracle of our modern era. It is packed to its organic gills with every conceivable type of food, all in impressive abundance. The cornucopia includes fresh French bread, humanely raised chicken, a dozen varieties of olive oil, wild salmon from Alaska, goat cheese from Switzerland, yoga magazines, wine galore, buffalo burgers, and an entire aisle dedicated to chips, salsas, and other snack foods. Today, I grab two apples, some organic dried apricots, a premade pesto-and-turkey sandwich, a bag of potato chips, and a cup of coffee to go. I'm in and out in under ten minutes. That's a miracle too.

Soon, I'm on the interstate, heading south. My mind drifts. The cornucopia in the natural foods store recalls a quote from the poet Ogden Nash that I read years ago. "Progress was good for a while," I think he said, "but then it went on and on."

Indeed.

Approaching the airport in Albuquerque, I sidle off the freeway and shake my head clear of road thoughts. It's time to concentrate. Airports are miracles too, though increasingly stressful ones. The flight is uneventful, and I arrive at my destination a few minutes early. Deplaning, I pick up my suitcase at a carousel, secure my rental car from a generic company (I can only tell them apart by their colors), and hit the road—all in under thirty minutes. That's amazing too, but it's all so familiar and routine to me by now that I don't pause to consider it.

I aim for the freeway and promptly make a wrong turn, though I don't realize it right away. When it finally occurs to me that I'm heading in the wrong direction, I punch the radio on. There's no rush—my appointments are not until tomorrow. As I drive, I dial the radio, discovering that it has nearly *two hundred* stations. Wow. It's satellite radio, which is still a novelty for me. Plowing through the stations one by one, however, I am quickly reminded of the modern cable TV dilemma: despite an explosion of choices, there's still nothing on. I settle on a classical music station.

My wrong-way drive brings a shoe store into view. I need a new pair of tennis shoes pretty badly, so I pull off the freeway and aim the car toward a cluster of outlet stores. Once inside, I am overwhelmed again by our economy's skill at *sheer volume*. The place is packed with shoes, which makes it feel like the organic grocery store, only with laces and socks. Drifting over to the men's section, I am confronted with at least seventy different variations on the basic tennis shoe. Making matters much worse, a friendly salesman tells me that it's two-for-one day. If I buy a pair of tennis shoes, I get another pair for free—any pair in the store. I look around. There must be six hundred pairs of shoes visible. I thank him, swallow hard, and start eeny-meenying my way through the inventory. Eventually, I emerge from the store dazed but triumphant, with two shoeboxes in a bag.

Back in the rental, I consult a map before driving to my hotel, which is conveniently located between the highway off-ramp and a large shopping mall. After checking in and depositing my belongings in the room, I drive over to the mall to explore supper options. The mall itself is ringed by chain restaurants, giving the impression that I'm entering an orbit around a giant, many-mooned planet. It certainly feels like a universe unto itself. Slipping beneath the outer ring of restaurants, I opt for a local sports-themed place on the planet's surface instead.

After a successful landing, I walk inside the restaurant, where I am immediately assaulted by a dozen very large television screens, each blaring a different sporting event. As I wait for a table, I scan the mammoth room, noting that every available space on the walls is occupied by something neon, mostly beer advertisements. I feel like I've walked into a holy place—the Temple of Brew. Observing my awed expression, one of the temple's acolytes approaches and guides me to a booth, where I plop down and dutifully order a beer from a very long list. I have no idea what I'm getting. I don't drink much, but I don't want to offend the beer gods either.

A burger later, I head back outside, where I decide to take a walk through the mall's mammoth parking lot. After a long day of sitting, I need the modest exercise. I didn't see a green park anyplace while driving, however, so I have to settle for the gray lot, which is sparsely populated, thankfully. I've heard that shopping malls are in trouble, victims of the Internet and changing consumer habits. This is a curiosity, personally, because it means the rise and fall of the mall took place during my lifetime. I remember the excitement the grand opening of a very big mall caused in my Phoenix neighborhood when I was fifteen or so. And talk about a temple! I loved wandering around in its air-conditioned comfort, gawking at the goods (and the girls). But if the shopping mall was another miracle, it was a short-lived one, apparently. Maybe we didn't make the right sacrifice to the mall gods.

It's a lovely evening. But the beer has made me droopy, so after three figure eights around the lot, I climb into the rental, drive back to the hotel, go to my room, grab a book from my bag, and slip into bed. There's no reason to turn on the TV, not even curiosity. No need to catch up on pop culture tonight. Besides, it's been a long, amazing day. *It's a remarkable world*, I think to myself as the god of drowsiness begins to work its magic.

Progress *is* good . . . *was* good . . . still is. I know it went on too long probably, but that's all right for the time being. Things are good—but for how much longer? The book slips in my hands. Despite our troubles, I feel fortunate to be alive today, now, here. I should give thanks to somebody, I think groggily. The U.S. Chamber of Commerce, perhaps, or my parents. Perhaps an offering to another god is in order—maybe the god of rental cars. Or central heating. Or fluffy pillows. The book slips again. I put it down.

I reach for the light and, with a twist of a button, darkness engulfs me once more.

TEN

A VIEW FROM EUROPE

(2010)

SUNDAY, SEPTEMBER 5, six a.m. A tiny breakfast nook, Imperial Inn B&B, Rome.

Gen, Sterling, Olivia, and I landed at Da Vinci airport at nine a.m. on Friday, tired and stiff after an uncomfortable, and for me mostly sleepless, flight from Chicago. We skipped the train into town, opting for a thrilling taxi ride to the B&B instead. That woke us up, if only momentarily. Arriving at our destination, we were somewhat surprised to find that the Imperial Inn resided in the corner of the fourth floor of a not-very-imperial building facing the busy Via del Viminale. I chose it because of its proximity to Rome's Termini subway station and because the online reviews were positive. Fortunately, the building had a quaint elevator that the kids loved, and the inn itself provided us with a room that featured two amenities instantly admired by the adults: a ceiling fan and soundproof windows. Rome is hot and noisy, as well as very large and overwhelming.

We couldn't wait to go exploring.

But not yet—everyone drooped from the long flight. Hunger ruled, so we hunted for a pizzeria, hoping to give our blood sugar levels a jolt. We found one around the corner, but the kids drooped all over their chairs, so we made quick, unauthorized decisions about pizza toppings. Fifty euros later, we were ready for a nap. Two hours later, we forced ourselves out of our beds and into the nearby subway, which took us directly to the ruin of the old Roman Colosseum in the heart of the ancient city. It was time to get this trip officially started.

Walking out of the dark subway station, we were awestruck by what we saw. The two-thousand-year-old Colosseum loomed, beautiful and terrifying at the same time. We stopped dead in our tracks. Gen and I were raised in Phoenix and Albuquerque, respectively, which means we grew up thinking that if something was built sixty years ago, it was *old*. Our attitude changed after we became archaeologists and our time horizon shifted backward by centuries, which helped us put the post–World War II development of the Southwest in perspective. Boom and bust, we came to see, were part of the cycle of things. But in Rome, we knew right away that our time horizon didn't go back far enough. To be confronted with a structure that was two thousand years old and five stories high was, well, a thrill.

Crossing the street, we entered the Colosseum quickly, thanks to a pass, and began to explore the structure, which had filled with the lovely orange light of late afternoon. It reminded us of a modern football stadium, down to the numbered entrances and bleacher seats (we imagined a Roman ticket: section XV, row XXIII, seat IV). It felt familiar and comfortable, at least until we began to consider the whole bloody gladiator thing. This was no college gridiron, of course. This was a killing ground, where many, many people died grisly deaths to the cheers of spectators. It was a chilling thought. Climbing a stairway, I recalled one of the videos Sterling and I watched before the trip, in which a British-accented narrator, after a particularly nasty episode of Roman military ruthlessness, said matter-of-factly: "The ancient Romans were not nice people."

Indeed, the Roman civilization poses a challenge to modern-day admirers, especially eleven-year-old ones. On the one hand, *what* Rome accomplished in its nearly one-thousand-year run is nothing short of stunning, particularly its engineering feats and countless victories on the battlefield. But those were just the headlines. What's always amazed me was Rome's ability to govern its far-flung empire as efficiently and peacefully for as long as it did. Sure, there were the occasional uprisings, and those pesky barbarians kept things hot on the frontiers for centuries, but in the main, the empire went calmly about its business. I knew this was a substantial achievement. Either the Romans were really good at governing, really lucky, or really ruthless. I suspect it was a combination of all three, with the "not nice" part playing a major role.

This brings me to the other hand: *how* Rome accomplished its indelible mark on world history. Much of it is a shocker for us moderns. Take the empire's heavy use of slavery, for instance, or the gladiator business, or the army's appalling practice of "decimation," in which a Roman commander enforced discipline by ordering every tenth soldier to be beaten to death by his fellow legionnaires. Slavery, however, is the tough one. Try explaining to your child why the Romans bought and sold human beings like cattle and treated them like dogs, especially when you learn, as we did, that one in four residents of the ancient city were slaves. Then there's the level of carnage that took place in the Colosseum itself. Olivia was very upset to learn that eleven thousand wild animals were slaughtered during one lengthy festival simply to satisfy the bloodlust of the crowd. *Eleven thousand animals!* Try making sense of that to a young girl whose favorite animal is a wild wolf.

For myself, the end-of-the-republic-start-of-the-monarchy thing that took place after five hundred years of republican rule is difficult too. One can't simply blame the dictatorial aspirations of Julius Caesar. Roman politics had become hopelessly gridlocked by the middle of the first century BCE, with the Senate blocking (sometimes violently) attempts at substantive reform. Elites and special interests dominated both politics and the economy, while discontent among the plebeians in the streets spread riotously. The future looked very uncertain, and tension rippled through every layer of Roman society. Eventually, civil war broke out, wracking the empire convulsively for a generation—and leading directly to the establishment of a hereditary emperor as Rome's new form of government.

I know the ancient Rome/modern America comparison is a cliché these days. But that's the thing about clichés—they have a ring of truth to them, or they wouldn't be clichés. As we wandered around the Colosseum, for example, awed by the way its mammoth blocks fit perfectly together and knowing that it was all accomplished by ropes, wooden lifts, and slave labor, I couldn't help but wonder about an analogy with our modern faith in technology. We have our own grand stadiums (sometimes called coliseums), of course, in all shapes and sizes, built primarily by the slave labor of fossil fuel. We keep building more and bigger stadiums too, heedless of the consequences, assuming that our ingenuity will overcome all obstacles, as the Romans undoubtedly believed as well. Unfortunately, technology hasn't helped us overcome the political gridlock in Washington, D.C.—spookily reminiscent of Rome, by the way. Is civil unrest in our future as well? Anyway, these thoughts were too heavy for the first day of a vacation. When the Colosseum closed, we headed for pasta and dessert.

It had been a very good day.

▪

THURSDAY, SEPTEMBER 9, six a.m. The breakfast room, Hotel Caneva, Venice.

I'm sipping a cappuccino from a machine and enjoying the meditative quiet of a Venetian morning. It's a big contrast with yesterday, which dawned with a spectacular rainstorm. Thunder boomed every thirty seconds or so, and rain fell in sheets. It was a portentous start to my fiftieth birthday, I thought. Actually, it was the second rainy day in a row, which is unusual for this time of year, or so the hotel clerk told me. I had gone downstairs to use a phone and saw that the rain had caused the adjacent canal to rise and flood the lobby. Wooden planks had been placed from the stairs to the dry part of the floor so guests could leave without getting their shoes wet. When I queried the clerk, he shook his head. "Very unusual," he said. "Usually in winter, not now." A quick flash of climate-change anxiety crossed my mind. Bigger and more frequent storms are an early sign of global warming, climatologists have been warning us. But then I thought, *Wait! I'm in Venice. I'm not going to think about that.*

We couldn't help, however, thinking about America. Over supper the previous night, Gen, the kids, and I did a quick comparison between the Old World and the new, and on many scores, America didn't fare so well. Take toilets. Here, you have two choices when you flush: small and medium. In America, you mostly have only one choice: large. That pretty much describes the difference between the two worlds. In Italy, people respect limits to their lives—street widths, room sizes, meal portions, the size of cars, the size of people. In America, limits are treated with contempt. This is one reason why Italy was such a pleasant surprise when Gen and I visited nearly two years earlier as delegates to the international slow food gathering in Turin. We were amazed by what we saw, including the way Italians accepted limits to their lives (except for restrictions on their driving habits).

We arrived in Venice on Monday afternoon without any trouble. The weather was clear, the trains ran on time, and the long walk from the station to the hotel, located near the Rialto Bridge, was, well, noisy. Our wheeled luggage clacked on the rough streets, echoing off walls of the houses, giving the underpopulated city an extra air of ghostliness. After depositing our bags in the austere but perfectly acceptable hotel, we headed down to St. Mark's square, where Gen and I quickly realized that Venice isn't really for kids, or at least doesn't hold the charms for them that Rome did. Venice is for lovers, of course, as well as for those who appreciate it being a "puddle of elegant decay," as the author of our guidebook put it. But it's not really for kids. The stores caught their attention, as did the pigeons, and they liked the labyrinthine feel of the city, which Sterling took as a challenge ("This way, Papa, I think"). But the subtle qualities of the city, especially the pleasures of texture that abound in Venice, were largely lost on them. That's okay—Venice has an elusive core that is hard even for adults to grasp.

Maybe it doesn't matter. When you walk into St. Mark's, it's the pleasure of texture that impresses. What a place! Even on a second visit, the space and the architecture and the light conspire to overwhelm the most jaded heart. It's a performance space too: the vendors, the dueling musical ensembles in front of elegant hotels, the swarm of tourists from every corner of the globe, and the countless pigeons. Sterling and Olivia began chasing the hapless birds through gaps in the crowds almost immediately. It was a timeless and endearing sight, I have to admit. I was proud to give them this opportunity—to chase pigeons around St. Mark's square. I think my birthday present came a few days early.

There was one disquieting moment. Leaving the square, we turned left at the waterfront and walked a short distance to the iconic Bridge of Sighs. We were stunned to see it wrapped in a splashy advertisement for a jewelry company. Later, we learned that the city had sold the space in order to raise funds to maintain Venice's vast, and deteriorating, cultural heritage. We also learned that the decision was highly controversial. I bet. America has its share of underfunded projects, but would it be okay if the National Park Service sold wall space at the Lincoln Memorial to a corporation for an ad? No.

Tuesday morning in Venice was drizzly, so the kids pulled out their homework. I went out for a long walk. It was the last day of my forties, and the drizzle fit my mood. What should I do with my fifties? I have been tilting at various Windmills for twenty years and now need to slow down. Being fifty means you are young enough to *want* to keep tilting at things even though your mind says you ought to know better. At forty, one can afford to be rather indiscriminate with one's battles, but now one must pick them more carefully. That's easier said than done, of course, but that's why you visit places like Venice with your kids—to look backward and forward simultaneously.

Late in the afternoon on Wednesday, we woke the kids from a nap and headed to a restaurant next to the Rialto Bridge that I had scouted. Incredibly, an empty table awaited us at the closest possible spot to the bridge itself. It felt like the supper gods were smiling on us. This was my dream, and it got better. An outstanding meal arrived. Gen ordered salted codfish with polenta, Olivia had pasta, I had pasta with seafood, and Sterling, who was indecisive for a change, picked two appetizers. Everything was delicious. We had bread, salad, and dessert. We talked and laughed and toasted. It was perfect. The night air was still and warm; the gondolas plied the Grand Canal quietly, dodging the water buses as the sky slipped slowly from dark blue to black. I couldn't have asked for a better birthday present.

I wished it would never end.

■

TUESDAY, SEPTEMBER 14, seven a.m. Hotel dining room, Beynac, Dordogne region of France.

I had just finished a lovely continental breakfast after a solitary predawn walk around the small medieval village of Beynac. We'd arrived late the previous evening, after a wrong turn in nearby Sarlat, and immediately fell in love with the town, impressed by the huge castle that rises directly above the hotel and the soft Dordogne River that flows at its foot. After checking in, we climbed a very narrow and steep street to the castle itself, huffing and puffing most of the way. We were rewarded with an amazing sight: a setting sun over rich farmlands embracing the tranquil Dordogne River—and another castle on the horizon. It was a balm to our weary tourist souls. That's because it had been a long two days getting there.

It began with toll road *hell*. We rented a car in Nice, but our guidebook didn't warn us about toll roads, and as a resident of the nearly toll-less American West, I had no clue what to do when I pulled up to the first booth other than panic. At the next booth, I inadvertently put the wrong ticket into an automated machine, which made it angry. I tried a credit card next, but it rejected it as American, probably out of spite. I didn't have enough coins either, so we decided to back up. Gen jumped out of the car and asked four miffed drivers to back up, employing her best college French. It was a mess. We navigated the next tollbooth without creating a crisis, but then we took the wrong fork in the highway, causing the attendant at the subsequent booth to squint at us with exasperation. It would have been comical if it hadn't been so stressful.

Eventually, the toll gods decided to stop tormenting us, and we made it to the Pont du Gard before closing time. This is the famous fragment of a two-thousand-year-old Roman aqueduct that carried water from the mountains to the ancient the city of Nîmes. It stretches spectacularly across the Gard River, three stories high and riddled with beautiful Roman arches. We arrived just in time to see the structure ablaze with fiery sunlight. I was awestruck by the aqueduct's perfect blend of form and function. It's an engineering marvel, of course, especially for its day, but it is also intensely beautiful, with or without the fiery sunset. "What an empire," I quipped to the kids as we strolled along the structure's base. I tried to think of something comparable in America, and the Golden Gate Bridge came immediately to mind, which was also constructed in an era where form and function meant something.

After a too-brief visit to the aqueduct, we hustled back to our car. We still had a long way to go and were eager to get to our next destination—the medieval city of Carcassonne—though the prospect of additional tollbooths along the way made my hands sweat. But the toll gods smiled on us. At the last booth, late at night on the edge of Carcassonne, we literally used up our last euros. If the toll had been one euro more . . . well, I don't know where we would have slept. That's because as we pulled up to our overnight destination, an old abbey, the gate was literally being closed for the evening. Two minutes later and we would have been searching the city for a place to lay our weary heads.

The next morning, I slipped out of the abbey and walked the short distance to the moated entrance of Carcassonne, which is a compact medieval castle and city surrounded by tall walls and filled with tourist shops. Dodging a steady stream of decidedly unmedieval delivery trucks, I wandered through the dampness, marveling at the moody architecture. Carcassonne was saved from oblivion by a mayor in the nineteenth century who had it restored to its former architectural glory just in time for the invention of the vacation. He did an admirable job—the place is a feast for the senses, as well as a hulking testament to the age of chivalry. As we prowled around the fortress after breakfast, we gawked at the huge scale of everything. The Romans were here first, of course—we could pick out their distinctive brickwork by now.

Later, while taking a long walk around the outer ring of walls, I marveled at the apparent lack of concern for personal safety on the part of the French. Frequently, nothing stood between us and a twenty-foot fall. In America, we would have been fenced in by handrails or ropes, especially if it were a state or national park. Looking in our faithful guidebook, I read a confirmation that the French don't worry about safety as much as we do. You can go largely where you want, it said. It is an interesting contrast. Sometimes it feels like you have more protection in America but less freedom.

At the approach of the noon hour, I directed us toward lunch in a pint-sized square in the center of town that I had discovered during my predawn ramble. It was now full of tables, waiters, dappled shade, and eaters. Three different restaurants shared the open space, which made for a fascinating dance of food, spirits, foreign languages, buzzing waiters, clinking glasses, and gorgeous light falling from the heavens. It was a feast for all the senses, the eyes especially. Later, we hit the gift shops and then elbowed our way back through the crowds to the drawbridge, where we lingered, soaking up the sights one last time. Tourism was a lot like the lunch we'd just had, I thought to myself: you eat a great meal in an exotic location, enjoying every bite, and then suddenly it's over. The glow fades, leaving you to wonder what the other dishes on the menu tasted like. You vow to return, snap a photograph, and move on.

And hope that the toll gods keep smiling.

▪

THURSDAY, SEPTEMBER 16, six a.m.

We were seriously hooked by Beynac and the Dordogne. We were due to pack up and drive to Paris, but I wasn't ready to leave. That's because I found a slice of Agraria here, and it had set me to thinking.

It started with our hotel and the lovely little village that surrounds it. Sandwiched between the river and the castle, with the countryside only a short walk away, Beynac strikes me as almost ideal (except for the constant flow of truck traffic just outside the hotel's front door).

It's a tourist town, to be sure, complete with art galleries and souvenir shops, but underneath the carefully maintained gloss is a real town that harmonizes perfectly with its surroundings. Take the "tabac" shop just up the street, for example. Owned by a very nice older couple who were amused by our daily purchase of exactly the same sandwiches, it is small, tidy, well worn, and full of things we needed. It has a "rightness" of scale that contributes to a sense of "placeness" that resonates throughout the village. Actually, we found this "placeness" everywhere we went.

We found it on Tuesday as we canoed the Dordogne River. After a morning of homework, we were picked up by the canoe company and driven to Carsac, where we were given two canoes and rather unceremoniously deposited on the riverbank by an unsmiling employee. It was a gorgeous day, warm and still. Sterling and I stepped carefully into one canoe, Gen and Olivia into the other. We shoved off into the wide, glassy river, knowing it would be a memorable day. For the next five hours, we paddled, drifted, dreamed, and oohed as we floated past limestone cliffs, pale tourists, old castles, farm fields, bridges, campgrounds, villages, and solitary houses. To call it an idyll would be an understatement, though the kids grew weary toward the end, and Gen's tender back began to hurt her from the exertion.

As we paddled, I thought: this area has it all—beauty, history, culture, farming, wilderness of a sort, fine food, civility, and a deep-rootedness that is literally foreign to me. It was an *agrarian* vision—a harmony of land and culture (without the historical exploitation of agricultural workers, I hope). Farmer and author Wendell Berry calls agrarianism "another way to live and think" and contrasts it to the destructive industrial model of living and thinking that has dominated the world for the past half century, especially in America. Agrarianism is "not so much a philosophy as a practice, an attitude, a loyalty and a passion," writes Berry, "all based in close connection with the land. It results in a sound local economy in which producers and consumers are neighbors and in which nature herself becomes the standard for work and production."

I had no idea, of course, if the agriculture we saw near Beynac was sustainable or not, but it looked like it to my eyes. There certainly was a strong sense of harmony to the landscape. It definitely gave me the impression that we were drifting through a slice of Agraria.

This feeling was reinforced the previous day by a drive to nearby Lascaux and a tour of the famous cave paintings of wild bison, horses, deer, and aurochs (cattle). Wherever we drove, we saw an enchanting mosaic of woodlands, small farmsteads, verdant grazing lands, picturesque villages, and healthy-looking riparian areas (the "wet zone" on both sides of a creek). We didn't see any overgrazing, obvious signs of erosion, or anything that looked like a feedlot. Maybe we were looking in the wrong places, but I doubt it. The area reminded me of the Amish country I've visited in central Ohio, only without the horses and buggies. This wasn't a coincidence—the Amish heartland is another slice of Agraria, a place where people have managed to live more or less harmoniously *with* the land, economically and ecologically, rather than against it.

This is one of the main lessons we learned from the existence of seventeen-thousand-year-old paintings at Lascaux. The cave's presence suggests that humans have managed to occupy this area for a very long time more or less sustainably—an observation that recalled the famous lament by the great American conservationist Aldo Leopold, who said that the "oldest task in human history is how to live on a piece of land without ruining it." Somehow, apparently, the residents of the Dordogne figured it out. How did they do it? Was it the soil? The rain? The area's remoteness? The culture? I knew that similar country had been ruined by hard use in other parts of the world. What was different here?

This is important because we live in an age where the issue of sustainability is becoming more and more critical, especially as the pace of "land ruination" picks up due to population pressures, energy demands, and food shortages. It behooves us, therefore, to study examples of sustainability in action, rather than in theory. However, to many American conservationists, the Dordogne may have little appeal. After all, there isn't much wilderness left there; it's all cultivated, to one degree or another, which is one reason why there are no wild bison, horses, or cattle around anymore. If it's wildness you're after, then rural France may not be the best place to look. But if it's harmony you're after, as Aldo Leopold was, then the Dordogne fits the bill.

And harmony is what I'm after.

I believe the twenty-first century will be dominated by the issue of declining human well-being, and in some places, it has already begun. If our ability to find food, fuel, fiber, water, and shelter becomes strained, then twentieth-century priorities such as wildness or endangered species protection will drop way down our to-do list. This is why it's important to find slices of Agraria. We need its harmony along with its food. We need to understand why human settlement persists so well in some places while in other places it has not. It's not simply a matter of rain, soil, climate, or other local factors; plenty of places with ample amounts of each have been ruined over time. The difference, I think, is culture—by which I mean our values, norms, and economic incentives. When they harmonize with the land, all flourishes; when they do not, despair follows.

That was the biggest lesson I learned on that trip: how our values shape our decisions. We can't simply make different decisions in the future (i.e., employ different technology) or "be smarter," as a popular corporate advertising campaign insists, in order to confront a crisis without confronting the value system that created the crisis in the first place. Whether it's the brutality of empires, the pleasures of texture, lessons about limitations, questions of form and function, or the importance of "placeness" in our lives, the view from Europe suggests that Americans need to ask themselves hard questions about values before we get much further down the road.

ELEVEN

LUCKY US

(2011)

APPARENTLY, *WE* ARE the future.

This thought crossed my mind last fall while waiting to turn left at a stoplight near my office. It was a gloriously bright day, and as I waited, I had the privilege to watch an elegant, old-timey convertible sail into the intersection, its top down, the wind blowing carelessly through the hair of its tanned occupants, both of whom appeared to be in their sixties. The guy behind the wheel sported a gray ponytail, and the woman seated next to him wore five or six silver bracelets on her right arm. From their casual but upscale clothes to the relaxed way their bodies seemed to be one with the convertible, the couple looked every bit the epitome of success, retired to the good life, and enjoying their presumably hard-earned leisure. As they crossed in front of me, I glanced down at the car's license plate, noting that the vehicle was from out of state, though I couldn't tell from where. What I didn't miss was the plate's vanity message. It read: LUCKYUS.

"I guess so," I said aloud as the convertible sailed out of sight.

The green left-turn light popped on. After navigating the turn, I eased into the slow lane of the busy boulevard as I normally do in order to let my fellow drivers, who always seem to be in a greater hurry than I, rush past. But this time, I let my truck coast a bit in the traffic, my foot barely pressing the accelerator, as I suddenly found it difficult to concentrate on the road. The image of the convertible and its relaxed couple was stuck in my mind's eye like a needle trapped in the groove of a vinyl LP record. Over and over they sailed nonchalantly through the intersection. LUCKYUS. LUCKYUS. I suddenly felt a surge of anger. LUCKYTHEM. UNLUCKYUS!

Recognizing a side street, I stepped on the brake and tugged the steering wheel sharply to the right, wanting badly to get away from all the hurrying and rushing going on in the world. A few moments later, I found myself pulling up in front of a small city park, whose emptiness and bucolic greenness offered a salve to my disquieted emotions. I knew this park. Gen and I lived near here for a while, back when Sterling and Olivia were infants, and we came to this little park often, usually with kids and a dog in tow. Today, however, my only neighbors were memories, which suited my mood just fine. I have a weakness for reminiscence, as Gen knows, especially when it involves our children. I decided to climb out of my truck to mingle with the ghosts.

I strolled the short distance to the play structure and sat on the end of the slide, my hands inside my coat against the sudden chill in the air. Staring at the neatly clipped grass and the wind-tossed leaves scattered here and there, the surge of anger returned. *UNLUCKYYOU*, I thought. *YOUKIDS*. I recalled a bumper sticker I saw on a mammoth gas-guzzling recreational vehicle one winter in Phoenix, years ago, while visiting my parents. It said: WE'RE SPENDING OUR CHILDREN'S INHERITANCE. I remember laughing. Funny!

No one is laughing now.

I kicked at the soft sand below the slide. Slowly, my anger dissipated into a sort of deep resignation. It's no longer a joke, all right. We *are* spending our children's inheritance, quickly too. The ironic humor of the bumper sticker was now officially painful. After all, I doubt this is what our parents and grandparents had in mind, exactly. Their generations *saved things*, for their families and for their country, so *our* lives could be better, more prosperous, more leisurely, and more enriching than theirs. They planned and saved for the misty future. And they did a damn good job of it too. Whether it was fighting a world war in defense of freedom, providing a quality education for their children, creating new national parks, or paying off a home loan, they tangibly enriched the lives of their descendants in a variety of important ways.

Our response, however, to all their planning and saving hasn't been as thoughtful. Basically, we looted their savings accounts. Cleaned them out. Sure, we did some good things with their money, such as send our own children to college, but a lot of it was squandered, burned up, or thrown away. And when the inheritance began to run low, we broke out the credit cards and began racking up huge debts, literally and metaphorically. We've been very selfish, frankly, and I think it would have saddened our grandparents. Not only did we fail to save for the no-longer-misty future; we carelessly indebted our children in the process. And now the bills, both large and small, are beginning to come due.

I kicked at the sand again and dug my hands deeper into the pockets of my coat. The air was still. The ghosts had all suddenly vanished.

▪

FOR MOST OF my adult life, I have worked as a conservationist, which means I have tried professionally to save things. When I was young, I campaigned to protect our national parks from a variety of threats, including air pollution, commercial development, and privatization. In the mid-1990s, as part of my new role as a Sierra Club activist, I went to Capitol Hill to lobby for the creation of new wilderness areas. I organized workshops on the ecological dangers of clear-cut logging on national forests and on the positive role that biodiversity plays in nature, and I helped the Sierra Club publish a citizen's guide on how to blunt the environmental damage caused by hard-rock mining in New Mexico. I led activist outings to Southern Utah, organized letter-writing campaigns, testified in public hearings, and fought a cynical, backdoor assault on environmental regulation at the time called "takings" legislation. I've written letters, columns, and op-eds, met with our Congressional delegation, and helped organize an effort to convince the federal government to purchase a privately owned ninety-thousand-acre ranch near Santa Fe and make it available to the public.

I tried my damnedest, in other words, to save things for the future.

This effort continued in 1997, when I cofounded the nonprofit Quivira Coalition. My conservation work took on a collaborative bent. I had learned a hard lesson from my Sierra Club experience that the missing piece of conservation was the positive role that *people* played. Environmental problems, I came to understand, were as much about *economics* as they were about the environment, thus requiring economic solutions to go along with ecological ones. I learned this by listening to the many heated confrontations between activists and ranchers and loggers over the years. Collaboration, not confrontation, was the key. Saving things, I saw, meant prudence, care, good stewardship, and positive relationships as much as it meant passing bills, creating new regulations, or establishing new parks. That's why I chose a quote from Wendell Berry as Quivira's unofficial motto: "We cannot save the land apart from the people; to save either, you must save both." Saving both became my mission.

In many ways, I had returned to the roots of conservation. The motivation to save, conserve, and use sustainably so that those who follow us may prosper is as old as the movement itself. And it was one of the chief reasons why conservation grew quickly into a national cause, embraced by Americans of all stripes, including many U.S. presidents:

> "I recognize the right and duty of this generation to develop and use our natural resources, but I do not recognize the right to waste them, or to rob by wasteful use, the generations that come after us."
>
> **—THEODORE ROOSEVELT, 1900**

> "The nation behaves well if it treats the natural resources as assets which it must turn over to the next generation, increased and not impaired in value."
>
> **—THEODORE ROOSEVELT, 1907**

> "We think of our land and water and human resources not as static and sterile possessions but as life-giving assets to be directed by wise provision for future days."
>
> **—FRANKLIN D. ROOSEVELT, 1935**

> "As we peer into society's future, we—and I, and our government—must avoid the impulse to live only for today, plundering for our own ease and convenience, the precious resources of tomorrow."
>
> **—DWIGHT EISENHOWER, 1961**

In another illustration, here are quotes from my three conservation heroes:

> "The major premise of civilization is that the attainments of one generation shall be available to the next . . . Civilization is not the enslavement of a stable and constant earth. It is a state of mutual and interdependent cooperation between human animals, other animals, plants, and soils, which may be disrupted at any moment by the failure of any of them . . . It thus becomes a matter of some importance, at least to ourselves, that our dominion, once gained, be self-perpetuating rather than self-destructive."
>
> **—ALDO LEOPOLD, 1937**

> "What a young American just coming of age confronts now is not a limitless potential, but developed power attended by destruction and depletion. Though we should have recognized the land as a living organism demanding care and stewardship, we have treated it as a warehouse, and now it is a warehouse half emptied."
>
> **—WALLACE STEGNER, 1981**

> "If we are serious about conservation . . . we are going to have to come up with competent, practical, at-home answers to the humblest human questions: How should we live? How should we keep house? How should we provide ourselves with food, clothing, shelter, heat, light, learning, amusement, rest? How, in short, ought we to use the world?"
>
> **—WENDELL BERRY, 1995**

Note the change in tone of these quotes over time. In the pre-World War II era, the tone was optimistic, though tinged with worry and warning. We still had time, most conservationists argued, to mend our ways and ensure a bright future for our children and grandchildren. This tone changes a lot, however, after the war as the American economy began to grow dramatically. Take Stegner's quote about America as a "half-emptied warehouse"—words that might have shocked Teddy Roosevelt in 1907, but by the mid-1980s shouldn't have shocked anyone paying attention. And that's before the real looting started as the global economy heated up. By the start of the twenty-first century, we were well on our way to emptying the planetary warehouse. So far, the answer to Berry's anguished question about how we should use the world is to *use it up*.

Somewhere along the way in the century from Teddy Roosevelt's day to the present, we, as a nation, slipped from the future tense to the present tense. We stopped using words such as "shall" and "will" and "must" and "tomorrow," replacing them with "now" and "today" and "mine." We saved things here and there—at least for the moment—but the principle activity of my generation, front- and back-enders combined, has been to take, take, take. It's as if we decided to shout at Teddy Roosevelt, "Thanks for the forests and the parks, we really appreciate them!" while waving a hand from the front seat of a sporty convertible, its top down, the wind blowing through the hair of its smiling occupants. "Thanks too for the cheap gasoline, the appliances, and the great coffee."

Somewhere along the way, in other words, we stopped thinking about our children. How else can we explain the colossal burden of resource depletion, financial indebtedness, and rising global temperatures that we are now officially bequeathing them? Of course, we didn't do these things *deliberately* to our kids. No parent would allow their child to be exposed to physical danger if they could help it. That's one of the main roles of a parent—to protect your children from harm. We're supposed to care for them, provide for them, and nurture them in ways that help them succeed in life, including raising them in a healthy environment. We're supposed to make their future brighter, not dimmer. Their lives should be better than ours. That's why we're supposed to save things, make personal sacrifices, and put their interests ahead of our own.

Right?

Wrong, apparently. We're not saving much of anything anymore, it seems. Oh, we're still trying, but under the specter of accelerating climate change, one has to wonder what we're actually saving in the long run. Take wildlife, for example, which has been the focus of so much conservation work over the last century. Have we really saved the whooping crane? The polar bear? The whales? The marmot? Tigers? Elephants? Is the future brighter or dimmer for these species? Dimmer, I think. And is it any less unethical of us to diminish the lives of plants and animals than it is to bequeath a diminished world to our children?

Perhaps more to the point: Did we save *us*? After all, it is becoming clear that we haven't conserved or saved things that are required for our well-being either. Take fossil fuels. Did we conserve them for future generations, stretching out a finite supply of these precious and irreplaceable minerals for centuries, or did we burn them up as fast as possible? What about the carbon dioxide content of the atmosphere—did we protect it from rising past a dangerous level, thus ensuring a climatically stable planet for our grandkids, or did we watch indifferently as totals rose? What about fresh water? Did we use it prudently so that the next three billion people to walk the planet will have adequate supplies of this life-giving liquid, or did we pump it like there was no tomorrow? The answers aren't clear yet. All I can say for certain is that somewhere along the way, tomorrow became today.

LUCKYUS.

▪

I KICKED AT the sand below the slide again, waiting for the anger to dissipate. I hate having these sorts of thoughts. I hate thinking that we've failed our children. I want to have hope and feel joy and work hard to save things for Sterling and Olivia and their friends. I know that I'm not free of guilt or blame, that I've done my fair share of consuming too. I've been one of the lucky ones as well. Still, I've tried to live modestly, carefully, and thoughtfully. And I tried to save things.

I dug my hands deeper into my coat pockets. Behind me, ghosts began to stir in the play structure. I could hear their laughter. Looking over my shoulder, I saw two-year-old Sterling climbing the stairs slowly, step by step, while his sister stood at the top of the stairs unsure about her next move. My nostalgia returned. Forget the future; I'd give anything to have my two-year-old twins back again, just for a few minutes. The wind picked up again. The ghosts dissolved. I stood and began to walk across the clipped grass and the tumbling leaves. At the truck, I paused. I turned to survey the little park again, this time letting the wind blow away all thoughts of children, past or future.

I suddenly realized that I needed to complete the chore that I had set out to do before becoming distracted by the old-timey convertible and the gray ponytail. I needed to get back to the office too, back to my day job of saving things. I'm not ready to call it quits, not by a long shot. I think there's still time.

Later, I looked up one of my favorite conservation quotes. It was written by Walter Lowdermilk, an American soil scientist who in the wake of the 1930s Dust Bowl was sent overseas by the U.S. Department of Agriculture to study the role of soil erosion in the downfall of ancient civilizations. His travels took him to Greece, Italy, North

Africa, Jordan, Iraq, and Israel. In all cases, he saw that the failure to save fertile soil from destructive overuse and erosion precipitated the decline of each culture. At the end of his tour, he composed his thoughts into an "Eleventh Commandment," which he read live in a broadcast over the radio in Jerusalem in 1939. Rereading it, I felt determined to renew my conservation work, determined to keep trying to save things:

> Thou shalt inherit the holy earth as a faithful steward, conserving its resources and productivity from generation to generation. Thou shalt safeguard thy fields from soil erosion, thy living waters from drying up, thy forests from desolation, and protect the hills from overgrazing by thy herds, that their descendants may have abundance forever. If any shall fail in this stewardship of the land, thy fruitful fields shall become sterile stony ground and wasting gullies, and thy descendants shall decrease and live in poverty or perish from off the face of the earth.

TWELVE

A NEW YORK INTERLUDE

(2011)

THERE'S SOMETHING ABOUT New York City that inspires the credulous.

At least, that's what it does to me. The first time it happened was in the spring of 2006, during an exploratory business trip to the Big Apple. Rummaging in an airport bookstore for something to read on the outward leg of my journey, I came across James Kunstler's best-selling cautionary tale about "peak oil" titled *The Long Emergency: Surviving the Converging Catastrophes of the Twenty-First Century.* Curious and alarmed, I plucked the book from the rack and flipped it over to survey the promotional blurbs, reading how the author "graphically depicts the horrific punishments that lie ahead for Americans for more than a century of sinful consumption and sprawling communities, fueled by the profligate use of cheap oil and gas." Yikes! Then I thought, "Oh, come on, how bad could things be?"

I handed the clerk fifteen dollars to find out.

Though I was initially skeptical, within a few pages I was hooked. Part of it was the subject matter, which I knew something about, but a lot of it was Kunstler's cheeky writing style, which engagingly mixed fact and speculation into a sardonic and irresistible stew of millennialism. As a conservationist, I was aware of various environmental calamities on the horizon, but Kunstler's book opened my eyes to the civilization-changing nature of inevitable oil depletion, as well as the interconnections of all our unsustainable activities, each grounded in our economy's messianic belief in unlimited growth on a finite planet. It all flowed into what he called the "long emergency," a sort of neo-collapse stage of civilization, replete with coffee shortages, oil hoarding, charismatic dictators, and a reduction to a pre-Model T standard of living. True or not, I couldn't put the book down. I read it avidly on the long flight, I read it as soon as I arrived at my hotel, and I read it between meetings. Eventually, I retreated to my room and kept reading until I reached the end, which required a marathon session in an overstuffed hotel chair.

Dazed and stiff, I decided to go for a long walk to digest all the bad news.

I started on the Upper West Side, where the hotel was located, and crossed quickly to Central Park. It was a beautiful spring day, with blue skies and blooming trees everywhere I looked—a perfect counterpoint to the gloomy vision of *The Long Emergency*. And yet, oddly, the blue skies and green trees reinforced the specter of societal collapse instead of challenging it. Strolling through Strawberry Fields, I imagined food riots among the foliage. I envisioned angry mobs gathering amidst the statuary literati of the park's south side, and overturned taxis at Fifty-seventh Street and Park Avenue, as I crossed the intersection. Macabre images, I knew, but also plausible, thanks to Mr. Kunstler.

Something like eight million people live in New York City—what would they do if the "elevators stopped working," as he put it, when electricity became scarce? How would eight million people be fed if trucks stopped coming or ships no longer sailed into port because oil production had entered its terminal decline? The mean streets of Manhattan—the streets that my feet were bruising at that very moment—topped out only feet above sea level. What would New Yorkers do when the Greenland and Antarctic ice sheets melted away, raising the sea globally by two hundred feet? What if the rains stopped falling on the city's watershed upstate, drying up the steel aqueducts that quenched millions?

As the bright trees of Central Park gave way to the brick canyons of mid-Manhattan, these scenarios drifted from plausible to inevitable in my mind. In fact, the "long emergency," I told myself, had already begun. The city's ceaseless traffic, as remorseless and irresistible as a gale, suddenly seemed fragile and impermanent; the hustling crowds looked tense and unsmiling, not from the pressures of big-city life, but from fear and desperation. The little food markets looked like shining jewels set in a drab urban mantle; the neon advertisements that competed everywhere for my attention with promises of beauty, harmony, and everlasting satisfaction began to look like false gods tempting us into sin and sloth at precisely the wrong moment in history.

Feet and mind aching, I decided to conclude my ramble at the site of the former World Trade Center. It seemed like a fitting metaphor for Kunstler's doom-and-gloom prognostications. After all, didn't another world as we knew it end on September 11, 2001, never to return? The site also served as a sober memorial of man's inhumanity to man. We are, unfortunately, a species capable of incredible madness and cruelty, evidenced not only by the gaping hole I saw before me but also in the knowledge of what lay in store for us if even half of what Kunstler predicted came true. If the world changed on 9/11, as I believe it did, then the chain reaction it represented was still evolving rapidly nearly five years later, though not in the direction that most Americans and their leaders supposed. We had set our shoulders against terrorism while ignoring far greater perils. We had also overlooked our role in bringing on these perils. We had met the enemy, as Pogo once observed, and it was us.

Clearly, Kunstler and New York City had messed with my mind.

During a return visit for business in November 2008, New York provoked my credulity again. This time, however, it involved not despair, but hope.

Two recent events had set the stage. The first was the ongoing meltdown of Wall Street. Not only did this event commence a historic economic and employment crisis in our country, but it also exposed our financial system's great swindle for all to see, and be outraged by. This generated the hopeful news that Wall Street's egregious behavior had created momentum for *real reform* of our financial system, finally. According to a wide variety of pundits, political leaders, and even bankers at the time, everything was going to be different now. American capitalism, we were told, had learned a sobering and corrective lesson and would change its ways—or have them changed by Congress. Inaction wasn't an option. Unregulated greed, lies, and shady dealings would soon feel the sting of governmental oversight (as they had not in the past eight years), just as they were now feeling the wrath of public outrage. Cries and promises of "never again" would be redeemed.

I believed it.

In my hotel room, I eagerly surveyed all the news channels, clicking rapidly through the promises, the denunciations, the shock, and the up-to-the-minute reports about the still-gyrating stock market. Dramatic stories about government bailouts, institutions "too big to fail," cratered pension funds, wipeouts of retirement plans, foreclosed homes, broken dreams, frustration, and fear dominated the news. I was mesmerized, appalled, and giddy all at the same time. Especially encouraging were the earnest calls for reform across the social and political spectrum. Many people in this nation understand that the status quo isn't working well, especially for the besieged middle class, and that an overhaul of our educational, financial, and free-spending governmental ways is now necessary. A megacrisis like the great swindle was an important opportunity, everyone said, to implement big changes.

I hoped so. Things certainly *felt* optimistic as I walked through the belly of the Wall Street beast to my meetings at various foundation offices. Although the people I encountered ebbing and flowing through the canyons of mid-Manhattan seemed to be as self-focused and impatient as ever, I thought the buildings looked embarrassed and contrite. They had an eyes-cast-down appearance, I thought, the tallest ones especially. For a lunch with an old colleague, I walked right into the epicenter of the great swindle itself. Entering the former headquarters of investment giant Lehman Brothers, whose collapse triggered the ensuing crisis, I could almost see the shockwaves emanating from the marble walls, radiating out like seismic lines on a map of America. I swore the buildings in this part of town leaned a little closer together, huddled in contrition, trying to apologize for the poor behavior of their occupants. A cold wind blew through the canyons now, adding to the dirgelike atmosphere of the moment—a moment, despite the chill, when I thought things really *could* be different.

The second hopeful event was the election of Barack Obama to the presidency of the United States. "Hope" and "change" had been the watchwords of his candidacy, and like millions of Americans who voted for him, I believed earnestly that he meant them. It wasn't just his words, however, that electrified our expectations. Not only had the nation elected its first African-American president, it had also tapped a serious, youthful, and obviously intelligent person to lead the nation, hopefully out of recession, debt, war, humiliation, and a host of other serious challenges. There was little doubt in my mind that Obama would implement his promises of change, especially since his party enjoyed large majorities in the U.S. Senate and House of Representatives. In the chilly wind, it felt like Christmas had come early—which present for America would the Democrats open first? Wall Street reform? Health care? Climate change? Immigration? That they would all be opened, sooner or later, was beyond doubt. I couldn't wait. I crossed my fingers.

Last month, I returned to New York after another two-and-a-half-year interval, again for business. My fingers were uncrossed. This time my credulity involved neither hope nor despair. Instead, I was taken in by the illusion of normalcy.

Five years after reading Kunstler's breathless opus, the peak oil crisis had not materialized. True, a barrel of crude oil was trading at over one hundred dollars—which had been unthinkable to industry analysts back in 2006—and true, the British Petroleum oil spill disaster in the Gulf of Mexico the previous summer had highlighted the dangerous extremes to which our addiction to oil had driven corporations and governments to act. But the much-promised chaos-inducing terminal decline of global oil production had not occurred, not yet anyway. The recession was an important factor, of course, since it had caused global oil demand to drop. High prices (and profits) had also spurred the exploration and development of new oil sources by corporations, leading to a "bump up" recently in oil production. New technology has created a glut of natural gas, driving down prices substantially, delaying "peak gas" as well, and leaving critics like Kunstler tapping their toes impatiently.

However, if the dire predictions of peak oilers haven't come true yet, it's a different story with climate change. In fact, the news from climate scientists has grown worse. Things are happening more quickly, and at accelerating rates, than even the worst-case scenarios predicted way back in 2007, when the Intergovernmental Panel on Climate Change (IPCC) issued its fourth cyclical report, warning of big trouble ahead. New and refined methods of observation, mountains of data, and a deeper understanding of the earth's paleoclimate record all pointed scientists toward the same conclusion: the climate situation is rapidly slipping beyond our control. Act soon to curb greenhouse gas emissions, scientists began to warn more vociferously, or face tragic consequences. The worrisome prognostications created a groundswell of media reports, meetings, conferences, and gatherings around the globe, culminating in the United Nations Summit at Copenhagen in December 2009.

The time for action, it seemed, was at hand.

In the end, nothing happened. The news cycle moved on, as did public interest. New leaders in Congress, elected last fall in a predictable and reactionary pushback against President Obama and the Democrats, took an avowedly hostile position on climate change. The possibility of meaningful action faded quickly, just as the science became more and more clear that inaction equaled tragedy. Meanwhile, the early effects of climate change are already being felt around the planet in the form of severe droughts, floods, and storms—just as the scientists predicted. Worse, the chaotic weather has strained the ability of food-producing nations, including the U.S., to keep up with the increasingly critical job of feeding seven billion people. Prices for staples such as wheat, corn, soy, and rice have skyrocketed in recent months and show no sign of relaxing anytime soon. The news cycle may have moved on, but if you live in Bangladesh, Somalia, Tunisia, or the Maldives, the crisis has not.

Wall Street certainly moved on. That's because the promised reform of the nation's financial markets never happened. Oh, in 2010 Congress managed, after much bickering between the two parties, to pass a tepid set of adjustments to the banking system, but few observers believe they will correct in any meaningful way the flaws in the way we do business as a nation. Instead, the big banks got bailed out by the U.S. Treasury with a slap on the wrist from Congress and a bit of finger-pointing from President Obama. In other words, Wall Street got away with murder. Official investigations never dug deep, no one responsible went to jail, real reform never happened, and as a result, American capitalism quickly returned to business as usual (where they are poised to repeat history). But, hey, the stock market bounced back! Profits are up. All is well, right?

The promise of Obama's leadership never happened either. Change and hope went begging. He did sign a law that substantially overhauled the nation's health-care system, but after that, Obama punted on nearly everything else. He picked a group of Wall Street insiders to run the Treasury Department and to advise him, thus sending a clear signal that his promise of "change" didn't apply to our financial system. He punted entirely on climate legislation, even as the Gulf of Mexico oil spill tossed an important opportunity into his lap. Worse, his administration has recently opened new fields for oil-and-gas and coal exploration across the nation's public lands, with more promised. He also expanded the war in Afghanistan and refused to close the detention facility in Guantanamo, contrary to campaign promises. Almost two and a half years into his presidency, it looks like business as usual still rules in Washington, D.C. Issued an invitation to the White House, Change never made it past the reception room. At least it did better than Hope, which got left on a street corner.

Visiting New York this time, and thinking of Obama, I was struck by the fierce *non*-urgency of now. Everywhere I went, everything looked perfectly normal. Tides of people pooled and flowed across intersections as usual; taxi cabs honked as urgently and noisily as ever; elevators worked perfectly; food and drink could be had at nearly every street corner; the neon advertisements continued to shout their utopian messages with deafening effect; tourists swarmed the theaters and toy stores in Times Square; and the trees in Central Park were as green and leafy as ever. The great industrial lungs of the city—steaming, sooty, and restless—breathed in and out at their usual relentless rate, oblivious or indifferent to any wider worry. Rain fell as it normally did, clouds scudded among the tops of the buildings, and the subways rumbled on. Walking endlessly as always, I easily believed that all was normal. Everything is fine. Despair, hope—what were those? Just words, echoes bouncing around the brick-and-steel walls of New York's canyons. Words fade, streets endure.

For a few days, I believed it.

THIRTEEN

THE WINDMILL

(2011)

I CAN RECALL the exact moment the Windmill struck.

It happened in late July 2008 as I stood in the lobby of the Nature Conservancy's Colorado headquarters in Boulder. An hour earlier, I had given a presentation to a small group of Nature Conservancy employees about ranching and conservation in the West—the subject matter of my book, *Revolution on the Range*, published two months prior. I was on a book tour, and the gathering had been arranged by a friend as a way of helping to get the word out. The previous evening, I spoke at a signing event at my favorite bookstore in Denver, also to a small audience. Feeling disappointed by both events, I lowered my guard as I entered the lobby after my presentation. That's when the Windmill struck, knocking me clean out of my saddle. It shouldn't have been a surprise. I knew that my idealistic desire to change the world ran the risk of a Don Quixote–like charge at the Status Quo, in the shape of a Windmill, but I had managed to avoid its whirling, indifferent blades up to this point. True, there had been a few near misses in the last year or so, but I had turned them into motivation for writing the book—to take the joust to a higher level. As a result, I knew I raised the risk of being struck, but I assumed I was up to the challenge.

I was wrong.

I staggered through the next few days, only regaining my senses while cooking breakfast for the family in a campground in Yellowstone National Park. We had scheduled the campout as an interlude in the book tour, a time to relax in our favorite national park and reintroduce Sterling and Olivia to its wonders, first experienced four years earlier. But in my daze, I had miscalculated the timing of the food. Breakfast was ready, but Gen and the kids were still fast sleep in the tent, so I sat down to a private meal of sausage and eggs, my stomach growling with hunger.

Seizing the moment, I decided to have a conversation with myself while I ate. *What was going on?* I knew I had been badly bruised by a blow, but rather than hitting solid ground as I'd expected, I felt myself sliding down an incline instead, heading for a dark place that was both distant and unfamiliar. *Where was I?* My body sat at a picnic table in Yellowstone, but my spirit resided in unexplored territory. Whatever the opposite of optimism is—fatalism, I suppose—I had found it. Or it had found me. Where was the exit, the path back? The food and fresh air, combined with a pot of coffee, had cleared my head somewhat, but when I urged my brain to scan for a map of this new terrain of sinking feelings, it came up empty.

My thoughts turned to the book. After two months of ho-hum sales and low general interest, it was beginning to look like *Revolution* would slip quietly into the Great Pond of Publication without causing much of a ripple. If true, this would be hard to take. It wasn't about money or praise or other conventional bruises to one's ego if your book doesn't sell well. No, my greatest fear didn't involve royalties or reviews. I wanted readers. Sales meant readers, and readers meant impact. I wanted the book to touch hearts and minds, but above all, I wanted it to *change things* in conservation and ranching—to improve the world and our prospects. It sounds cliché, but that was my dream. Of course, having this dream meant tilting at the Windmill. It was worth the risk, I thought. What had happened to the man from La Mancha wouldn't happen to me, right?

I took a long sip of the rapidly cooling coffee and looked across the campground as early morning sunlight sliced through the lodgepole pines. Here and there, people stirred at their tents or trucks or stoves. Although the day promised to be warm, wood smoke drifted across the land like a comforting shroud. Through the trees, I could see the outline of National Park Mountain, so called because it rose above the meadow where, in 1870, a party of explorers camped after a soul-stirring sojourn among geysers and waterfalls. According to legend, as they sat around the campfire that final night, one of them broached the idea of a national park for the area. It was a way, he said, of forestalling the pell-mell rush of developers and speculators that was sure to follow in the wake of their exploratory report. It was a radical idea for the time, but one that quickly resonated with politicians and the public alike—foreshadowing both the rise of the American conservation movement and the devastating exploitation of the natural world that would take place in the twentieth century.

Less than two years after the fateful campfire, President Ulysses S. Grant signed a bill creating Yellowstone National Park, the first in the world. Change, in other words—big change—began only yards away from where I sat sipping coffee. It wasn't a dream.

I studied the sunlight on the mountain for a while. Feeling calm for the first time since Boulder, I cautiously asked myself a question: *What had I intended to change?*

I thought back. The book had begun as a series of profiles of ranchers, conservationists, scientists, and other westerners I had met who were doing fabulous, progressive, collaborative things for the land and its people. The process of learning their stories had been a revelation to me. Raised in the environmental movement, I had been told repeatedly by my fellow Sierra Club activists that extractive use of the land, especially logging and livestock grazing, was a zero-sum battle—meaning that conservation could only advance as far as chainsaws or cattle retreated. I believed them at first, but then my college-trained skepticism of orthodoxy kicked in, and I began to ask questions. Was it really an either/or situation? Was there no win-win for the environment and local economies? And why, I wondered, did every solution to an environmental ill proposed by activists always seem to carry the maximum penalty for rural people?

In 1996, for instance, the Sierra Club adopted a national policy that opposed all logging on public lands, even small-scale projects. Dubbed "Zero Cut" by its supporters (and its critics), this policy quickly caused a stir, especially among loggers and other members of the working rural poor who correctly saw themselves as bearing the brunt of its effect. It was not a surprise, therefore, when Hispanic woodcutters in traditional villages across northern New Mexico vigorously protested the Sierra Club's new policy as racially discriminatory, economically destructive, and anti-rural. Things came to a boil when a group of Hispanic loggers hung two prominent environmentalists in effigy in front of the state capitol building one sunny day. Their anger and criticism caused a great deal of hand-wringing among Sierra Club members, both in New Mexico and nationally, as I witnessed firsthand.

Quickly, my questions turned into objections: Why a ban on all logging? Why hurt poor people? Isn't there a family-scale, sustainable way to cut trees in a forest? Don't most scientists insist that our forests are overgrown and in need of thinning for their health? Can't we find a way to employ local people in this work?

Speaking up with these objections in print, I received a series of nasty rebukes from my fellow activists. One cornered me at a meeting and accused me of "caring more for people than the environment." Another activist publicly accused me of being a stooge for the wood pulp industry. Another tried to get me evicted from the Club.

Clearly, I had angered a Windmill.

After a pause to sort out my thoughts, I decided to keep asking questions, this time about cattle. I had met a rancher who did things differently; he moved his cows around in a way that mimicked the behavior of wild herbivores and insisted that cattle and wildlife could get along. These were great ideas, I thought, and so we decided to explore them by founding the Quivira Coalition. In the process, I stepped away from the long-running brawl between ranchers and environmentalists and looked around. What I discovered was a great deal of progressive, collaborative, and regenerative work taking place across the American West.

This was news to me, and I quickly realized why—it wasn't being reported in the media. So in early 2004, I approached the editor of an online daily news service that focused on the Rocky Mountains, proposing to write an irregular column about this "other West" that I had discovered. He agreed, and I began crafting profiles. Two years later, I retreated to a cozy A-frame cabin on the James Ranch, north of Durango, Colorado, to knead the essays into a book. Working by intuition, I shaped a loaf out of my scribblings and then sent it on to a colleague for his opinion. His advice came quickly: send it to a publisher he knew. I did so, and to my amazement, they accepted it, assigned an editor, sent me a contract, and set a tentative date for publication. The whole episode blew past in a blur. I was going to have a book published! Readers awaited.

I took another long sip of coffee, watching the sunlight on the mountain. A slight breeze picked up, ruffling the branches above my head. I glanced into the tent, but Gen and the kids were still asleep.

What was I trying to change with the book exactly? Many things: the sour relationship between ranchers and environmentalists, poor grazing practices, the trouble with us-versus-them thinking, conflicting ideas about use of natural resources, persistent urban-versus-rural divides—a lot of different Windmills, in other words, with a lot of whirling blades, but eminently chargeable from the same horse with the same lance, I was certain. Apparently, I was wrong. There weren't a lot of little Windmills; there was just one—the one that struck me in Boulder. It goes by various names—the Status Quo, Business as Usual, Tradition—and it was far more invincible than I had ever imagined. In fact, at that moment, sitting at the picnic table in Yellowstone, staring at sunlight, it looked downright monolithic and unyielding.

I cleaned up the remains of breakfast, refilled my coffee mug, and decided to go for a walk along the nearby Madison River. Shortly, I came to a spot where a small side channel of the river cut a curve in the bank. Looking down, I saw concentric circles of stones in the water, marking the location of hot pools. A memory came rushing back. I knew this place. I had soaked in one of these pools. Teleported back to the summer of 1977, I was sixteen years old again, visiting Yellowstone as part of a backpacking odyssey through a series of national parks that culminated with a hike on the John Muir Trail in California, named after the Sierra Club's famous founder. It was a life-changing experience, and when I returned home, I embarked on a lifelong adventure in conservation. It felt like a million years ago.

The steam mingled with a thin mist above the pools, giving the scene an ethereal feel, like a postcard or a scene from a travel documentary. I lowered myself carefully to the riverbank, not wanting to spill the coffee, and sat on the cool dirt. I studied the mist, knowing that when the sunlight reached the vapors they would quickly dissipate. I took a sip from the mug. *Sixteen*. Yellowstone blew my adolescent mind. Three memories stood out: sitting on the front steps of the General Store at Old Faithful eating an ice cream cone while watching a fascinating parade of fellow visitors ebb and flow around me; hiking past smelly Turbid Lake at the start of our four-day backpacking trip into the upper Lamar Valley; and soaking in one of these hot pools upon our return, worn out but exhilarated.

1977. That *was* a million years ago. No computers, no cell phones, no Internet, no 24/7. Just disco music and Jimmy Carter. Social networking meant going for a hike with friends or attending a party, not like today when it means sitting alone in front of your laptop "friending" other people who sit alone in front of their computers. Isolated, controlled, tethered, safe, inside, and virtual—it's become a world far different from the one I knew when I soaked here all those years ago. The Windmill didn't exist then either, at least not to me, though it was likely under construction. Certainly, the blades had not been attached yet. When had it become fully operational? 1990? 2000? Was my quixotic adventure doomed from the start?

Feeling blue suddenly, I rose to my feet and headed back to our campsite. My head had cleared, but a mist had settled on my heart.

For the next two days I went through the motions of Yellowstone, visiting visitor centers, inspecting geysers, eating hamburgers, and driving endless miles. The mist lifted a bit when the kids completed the Junior Ranger program, earning a coveted plastic badge, but it descended again when we lost my expensive digital camera, left accidently on a bench near a bubbling mud pot. We dashed back, but it was gone. We went for ice cream cones at the General Store as a consolation.

The next day, the mist lifted unexpectedly. It happened during supper at Roosevelt Lodge, located in the less-visited northern portion of Yellowstone. We had driven over the hill into this part of the park in search of wildlife, wolves specifically. Olivia had developed a passion for wolves and desired to see one in real life. Sterling preferred moose. A kindly ranger told us that both could be found, maybe, near the Northeast entrance, so off we went. We saw crowds of bison in the lower Lamar Valley, but no wolves, though we did spy a moose at a distance. No one was disappointed, however. It had been a beautiful day with loads of fresh air and great views. Wildfires were burning someplace in the park, casting woodsy smells into the wind, which seemed appropriate since it was the twentieth anniversary of the Great Fires of '88, when a huge portion of Yellowstone burned up. We saw the aftermath everywhere we went—carpets of twenty-year-old lodgepole pine forest, demonstrating once more that nothing in nature is static, even in a national park.

Feeling hungry at the end of our long day of driving and looking, and not much in the mood for the hot dogs waiting for us back at the campground, we decided to have supper at the historic Roosevelt Lodge, near Tower Junction. We arrived just in time to watch the sunset from the Lodge's magnificent wooden porch, whose rocking chairs beckoned invitingly. This was my first visit to the old lodge. Built in 1903, the building exuded a "woodsiness" that immediately appealed to something deep inside me. It wasn't simply nostalgia for a lost era or a romantic impulse brought on by the rocking chairs; it was something else, something that was missing in our helter-skelter, plugged-in, overloaded, and stressed-out modern world that I couldn't put my finger on right away.

Perplexed, I went inside and ordered a round of Shirley Temples.

We settled in on the great porch along with a dozen other visitors and soaked up the remains of the day. I rocked for a while silently, determined to think of nothing in particular when the "something is missing" feeling returned. It happened as I caught snippets of conversation among our neighbors in their chairs, chit-chat about recent fishing adventures, brushes with wildlife, next destinations, the weather, and so on. After a short while, their comforting words blended together. The soft sounds of their voices and laughter wrapped me in a blanket of companionship, warming my heart. I closed my eyes as the daylight slipped into evening glow, and I felt the mist begin to lift.

By the end of supper, it had evaporated entirely. The comforting blanket of human voices gave way to an energetic vitality as the four of us talked and laughed our way through hearty meals of BBQ pork, spaghetti, and hamburgers. We told stories, recounted adventures, made plans, and acted silly. The world shrank to a circle the size of a dining room, filled with strength and joy and love, warming my bruised heart, melting the mist. It wasn't just Gen, Sterling, and Olivia—the whole room, which was full of people and their happy noise, had the feel of a large family gathering, as if we had all just attended a really great wedding. Although we were all complete strangers, undoubtedly representing a wide spectrum of states and nations, for a brief moment, we were united by food, drink, and conversation. As I listened to the laughter and the clink of glasses, I felt my wound begin to heal a bit. I felt like I could move my arms again and, perhaps, if I wanted, stand on my feet. The shock of the blow still reverberated, however, so I decided to take things slow and try to steer clear of the Windmill for a while.

▪

I RETURNED TO Yellowstone exactly a year later. I needed to. It had been a rough twelve months on the save-the-world front. Although I had kept my vow to leave my lance on the ground, the Windmill still managed to lacerate my optimism, repeatedly reopening the wound from its previous blow. It ached painfully, despite my attempts at self-doctoring, so I thought another visit to Yellowstone's healing waters would lift my spirits again. Gen had recently started a new job and had not yet earned enough vacation time for a long trip, so I packed up Sterling and Olivia, announcing to them that we were off on a new search for wolves and moose, kissed Gen, jumped in the truck, pointed it north, and hit the road.

I secretly crossed my fingers, hoping that somewhere among the geysers and mud pots I would find the bottom of the trench I had fallen into. The emotional slide that began with the Windmill's blow in 2008 had accelerated over the year as I watched the Status Quo prevail in a number of dispiriting ways, especially at my day job on the ranching and conservation front, where I absorbed two big disappointments.

The first was the idea of a "grassbank," which is a physical place where grass can be exchanged for conservation work via cattle grazing—the subject of a chapter in my book. Beginning in 2004, the Quivira Coalition had directed the Valle Grande Grassbank, located near Santa Fe, making it one of a dozen similar projects across the West. Grassbanks bucked Business as Usual, especially on public land, with their collaborative and progressive goals, injecting innovation into a sclerotic federal bureaucracy as well as provoking the ranching and conservation communities to rethink paradigms. Early signs were hopeful. An introductory conference in the year 2000 drew hundreds of people. Grassbanks popped up all over the West. There was even talk about starting a National Grassbank Network.

In the end, however, the Windmill would not be denied. By 2009, all grassbanks except one on private land in Montana had fizzled, including ours. Enthusiasm waned as bureaucratic, cultural, and financial obstacles refused to budge. Partners dropped out. Bills stacked up. To make ends meet, we began to run cattle on a portion of the Valle Grande Grassbank, but even this fell apart in the fall of 2008 when the economy went south, forcing us to sell our herd. Eventually we decided to sell the ranch altogether, ending our grassbank experiment for good.

The other letdown involved the Valles Caldera National Preserve, a ninety-five-thousand-acre privately-owned ranch near Los Alamos, New Mexico. After much wrangling, it had been purchased by the U.S. government and given the novel mandate by Congress to break even financially while (1) maintaining itself as a working cattle ranch,

(2) providing recreational opportunities for the public, and (3) preserving its natural beauty. It was a tall order, to be sure. In fact, it was supposed to be a new model of federal land management, one that broke through the deep paralysis on public lands by blending conservation and natural-resource use into a sustainable whole.

It was an exciting idea, and we were pleased when Quivira was chosen to be a part of the grazing team on the preserve in 2007. We promised both sound stewardship and a financial return to the American taxpayer for the privilege to run livestock sustainably in such a beautiful and productive landscape. Our effort was successful too—but it fell on deaf ears. In the following year, the cattle management returned to Business as Usual, and by 2009, it had become clear that the entire management of the Valles Caldera was slipping into Status Quo mode and would not be providing a new model at all. Rumors circulated that legislation would be introduced in Congress to turn the preserve into another national park (a cherished dream of environmentalists). If true, it meant that a golden opportunity for reform had slipped away.

It all added up to a big desire to return to the rocking chairs at Roosevelt and order another round of Shirley Temples.

There was another reason to go back to Yellowstone, however, that did not involve headlines, politicians, ranchers, bureaucrats, or book sales. It did involve tilting at a Windmill, but not the one that had struck me in Boulder. This one was different, though no less implacable. Instead of representing a resistance to change, it stood for the impossibility of slowing change down. Gen and I had built it ourselves, at our house, where the whoosh from its blades grew louder by the day.

Our children were growing up.

At ten and a half, Sterling and Olivia were fast approaching the point of no return for childhood. I wasn't ready. I didn't want them to leave the land of Honah Lee yet, abandoning poor ol' Puff the Magic Dragon to his cave. And why was it happening *now*, just as my saving-the-world spirits had sunk? It didn't seem fair. In weak moments, a bad case of sentimentality would come over me as memories of the kids at five, the trips we took together in those heady early Quivira days, and our adventures of discovery tumbled together in my mind unproductively. I loved every minute of our time together as a family, and I wanted it to go on and on, knowing perfectly well it was impossible. Hardest of all was watching the kids grow out of their wonder and amazement—out of the world of Harry Potter wands, pirate battles, LEGO fortresses, fairy wings, and unicorns. For ten years, Gen and I had the deep privilege of sharing and nurturing that wonder and marveling at its effects. Lately, however, Puff had started to feel lonely as Sterling and Olivia came around less often, making him sad. Apparently, they had new adventures to pursue in a more grown-up land. One day, of course, they would not return at all.

I was on Puff's side. So, if nothing else, heading back to Yellowstone meant I could hit the pause button for a week or so in Sterling's and Olivia's fast-evolving lives, make some new memories, and enjoy their gosling-ness for while longer.

We entered the park via its Northeast entrance. It had been a long day of driving, so we zoomed right to Roosevelt Lodge, vowing to stop only if a moose or a wolf actually trotted across the road in front of us. Once we were settled into our rustic cabin, I headed for the lodge's big wooden porch and the rocking chairs. The moment my feet went up on the railing, the peaceful feeling I had sought for months descended upon me like a spring shower. It was good to be back.

After a short while, Sterling and Olivia wandered over, knowing perfectly well where I could be found. They settled into adjacent chairs, rocking quietly. Each had brought a book, and a stuffed animal. I asked about appetites, but they shook their heads no. They wanted to read. I rocked idly for a while, thinking. It had been a long journey to get here, miles and months and years and diapers and dollars and worries and memories—all hard earned and all too quickly left behind. Something didn't seem quite fair about it. The kids kept reading, so I decided to watch people come and go. Cars came and went.

Time rolled on.

The next day, an amazing thing happened—we saw a pack of wolves. Sunning themselves on a hillside, they were the essence of canid cool, stretching lazily in the bright light, completely ignoring the horde of gawkers and well-wishers who lined the road nearby. One watcher let Sterling and Olivia peer through a spotting scope, which made their day. Mine too. I looked briefly, not wanting to take any precious time away from the kids. They looked and looked. The thrill subsided after a while, pushed along by the nonchalant attitude of the wolves. No matter, the discovery had been made, a threshold crossed. We had seen wolves! We thanked the owner of the scope and pushed on to Mammoth Hot Springs, where Sterling and Olivia made a beeline for the gift store. I knew what they wanted: more stuffed animals. One wolf and one moose. I happily pulled out my wallet.

For the next six days, I reveled in parental heaven. We:

Swam in the Firehole River.
Read books in a great hall.
Waded to a log in a river.
Laughed at waves on a lake shore.
Listened to Marty Robbins sing Western ballads.
Piloted a boat across Yellowstone Lake.
Held our noses among stinky mud pots.
Hiked to a waterfall by moonlight.
Talked to a bison.
Strolled through a meadow.
Watched a movie about volcanoes.
Gave each other lots of hugs.

On the morning of our last day, I rose early and went for a predawn stroll among the geysers in the Old Faithful Basin. Not wanting to leave, we had decided to stay an extra day, but Roosevelt was sold out, so we shifted to a cabin near the Old Faithful Inn—which was both lucky and a blessing. It was a lovely spot, right next to the murmuring river and a stone's throw from the great geyser itself. Enchanted, we walked a big loop through the basin the previous evening, timing it so we could catch Old Faithful going off just as we approached from a less-crowded direction. Later, after a supper of spaghetti and garlic bread in the cafeteria, we wandered out again to watch the eruption once more, this time in the bewitching light of late evening. We stood in silence for a long time in front of the sublime sight, marveling at the power and mystery of nature. I held Sterling's and Olivia's hands.

My predawn stroll the next morning was shrouded in fog. A thick, cool mist had settled on the basin, giving it an otherworldly feel, as if I were walking on the moon of a strange planet (though with information signs to read). The wooden walkways were perfectly discernable, however, so I relaxed as I walked and let my mind wander in the mist. How did I feel? Better. The sliding feeling that had started a year earlier was gone. I stood on solid ground. This was good. However, tall walls loomed all around, telling me that I was still standing in something deep. And I could hear the whoosh of the Windmill's blades far above me, turning relentlessly. Climbing out of wherever

I had fallen would be difficult, I knew, fraught with the risk of more blows. Staying down here wasn't an option, however. For my sanity's sake, I had to find a way out.

I saw a way. I perceived it on our last night at Roosevelt. Walking back to our cabin after a lovely supper together, Sterling and Olivia suddenly bolted to the top of a small hill, feeling spunky. I watched them as they horsed around in the amber light. They did a little dance together, spontaneous and childish, as if they were listening to a happy tune that was beyond the range of my adult ears. I envied their ability to hear it, and to pirouette about like that, so carefree. For a moment, the wound where the Windmill's blade had struck began to throb painfully, as a surge of sentimentality rose. I missed the early years—and the optimism they represented—a lot. Back when things were fresh.

I pushed it all away. I was holding onto things that needed to be let go, I realized, holding them fast to my chest, making myself sick. It wasn't just about my goslings. My health, I saw, was up to *me*—not to news headlines or Congress or bureaucrats or stuck-in-the-mud ranchers and conservationists. None of them had a cure for what ailed me—only I did. The first step was to release things that I couldn't control and let them sail away on an evening's breeze, like music I could no longer hear.

My heart ached suddenly, there on a walkway, in the fog. Letting things go—childhood most of all—is a lot easier said than done, especially when the future looks so uncertain. Not wanting to walk any farther, I detoured into an educational pullout and gazed down at the sign in front of me, covered with incomprehensible symbols. I think I stood in front of another geyser, but my eyes weren't focusing properly. Looking up, I thought I could detect the outline of the Windmill among the vapors, spinning silently. I closed my eyes. The Windmill wasn't going away, I realized, not now and probably not ever. Did I want to find a new steed, climb on, and continue my quest? *Yes.* I wanted to continue. For Sterling and Olivia, I needed to keep trying. I needed to push on, explore new country—carbon country, as it turned out. To get there, I knew I would need a new map. I'd begin right away, I decided.

I opened my eyes. The fog had thinned a bit, and I could tell that dawn wasn't far off. The hour had flown away. It was time to head back to the cabin, rouse my fledglings, and head home. I hesitated, however, in the turnout. Pivoting here meant taking my finger off the pause button. Sterling and Olivia would commence growing again, pushing on rapidly to adolescence, adulthood, careers, relationships, bills, and their own Windmills. I wasn't ready. I leaned on a railing, peering into the past. Puff was out there somewhere, in the gray-white mist, his great-horned head resting nobly and sadly on the floor of his cave, waiting. Words of comfort were useless. He knew the rap, and was inconsolable.

I was consoled, however, by the light, the fresh air, and the love I knew was waiting for me. I felt better, much better. I decided: you do what you can, you push on, you pray, you light a candle, and You give hugs to friends and family. Be a positive force, have a heart, and don't forget the music.

Time is the great leveler, poets tell us, the ultimate arbiter of our fates. It rolls on as relentless and unfeeling as erosion, and as mighty as a thunderstorm. Tilting at time's merciless current, I now understand, is like trying to keep your children from growing up. You can hold on for a while, but eventually you have to let go.

Just ask Puff.

FOURTEEN

CHASING IRENE

(2011)

"Don't ask yourself what the world needs. Ask yourself what makes you come alive and then go do that. Because what the world needs is people who have come alive."

—HOWARD THURMAN, CIVIL RIGHTS LEADER

WHAT MAKES US come alive?

Danger, for starters. I certainly perked up when I learned that my plane was scheduled to land at Philadelphia's airport at the very hour that Hurricane Irene was due to strike the city. Exactly six years earlier, Hurricane Katrina devastated New Orleans in the costliest natural disaster in American history, not to mention the eighteen hundred people who died, and though I knew Hurricane Irene wasn't in the same league, I didn't feel like taking a chance. So I called up the airline and received permission to postpone my flight by twenty-four hours. Thankfully, as it turned out, not simply because I was safe, but by traveling in the hurricane's wake for the next seven days, I had an unexpected opportunity to explore the things that make us come alive and contemplate what this might mean as we move further into the Age of Consequences.

Born in the mid-Atlantic on August 20, 2011, Irene developed into a Category 3 hurricane as it moved west through the Bahamas.

Although it lost strength as it veered north, the storm played a nerve-wracking game of nip and tuck along the Eastern Seaboard, forcing massive evacuations and countless business closings. In Washington, D.C., a major ceremony inaugurating a memorial to Martin Luther King Jr. had to be postponed. In New Jersey, the storm caused Atlantic City's casinos to shut down for only the third time in their history (the last hurricane to strike the Jersey Shore was in 1903). By the time Irene's eye passed over New York's Central Park, it had struck a patch of dry air and weakened to merely a tropical storm. It drenched the city but didn't cause substantial damage. To New Yorkers, it felt like a bullet dodged. In typical New York style, one resident later quipped that Irene was "just another storm."

It certainly wasn't "just another storm" to residents of Vermont, however. Parts of the state were pummeled with as much as eleven inches of rain, turning mild-mannered streams into raging torrents that knocked homes off their foundations and took giant bites out of roads. Cemeteries flooded, with caskets floating away. Three people died. Vermont Governor Peter Shumlin called it the worst flooding in a century. It was the same in Upstate New York, which received over thirteen inches of rain from the storm. Gale-force winds flattened trees and toppled telephone poles like matchsticks. "We were expecting flooding; we weren't expecting devastation," said one resident. "It looks like somebody set a bomb off."

By the time Irene dissipated its energies over northern Maine, it had carved a trail of destruction from Florida to Canada. At least fifty-five deaths were blamed on the storm. Approximately seven million homes and businesses lost electrical power. Extensive damage occurred along coastlines as a result of storm surge. Tornadoes spawned by Irene caused significant property damage. Dozens of rivers, spurred by a rainy summer that had saturated soils, reached one-hundred-year flood levels. One river, in Greene County, New York, reportedly reached the five-hundred-year flood mark. The list went on and on. Ultimately, the price tag for Irene's damage would reach $15 billion, making it the sixth-costliest hurricane in U.S. history.

Despite the damage and the cost and the lives lost, however, there was a general sense that things could have been much worse. Flying into Philadelphia the day after the hurricane, under azure skies, the only visible sign of distress that I could see from the air was flooding along the Delaware River. It didn't look terribly bad—and it wasn't. Taking a solid punch on the chin from Irene, Philadelphia had barely staggered. Sure, there were a lot of downed trees and electrical lines, as well as substantial water damage, but by the time the wheels of my plane touched down on the sunny tarmac, repair operations were in full swing. Humans were on the job. It would have been different, of course, if Irene had remained a Category 3 storm. Nevertheless, as the plane taxied up to the gate and I watched the beehive of activity all along the airport terminal, I thought, with some pride, about the amazing way humans rise to the challenge of emergencies. We're fabulous short-term problem solvers, and things like floods and hurricanes bring out the can-do side of our nature.

Emergencies make us come alive.

This is a good thing. If even half of what scientists predict about the future comes true, then crisis management will become a way of life for us, requiring as much can-do spirit as we can muster. I have no doubt we'll tackle each crisis with skill and gusto.

This oddly hopeful feeling was reinforced as I drove north out of the city. Fields of twisted and flattened corn, piles of shattered tree limbs, and pools of floodwater at creek crossings dotted the countryside, as if a gang of giants had stampeded across the land in a fit of chaotic anger. The blue skies overhead only added to the surreal effect. Still, evidence of recovery abounded. I saw crews repairing downed electrical lines and trucks hauling debris away. There was also heavy traffic on the roads—as if nothing unusual had just happened. It was another positive sign, I thought. For once, I was *glad* that a truck rode my bumper.

I had flown to Pennsylvania to visit the Rodale Institute, near Kutztown. Rodale is a well-known center of organic farming research and education, and I wanted to interview Jeff Moyer, the longtime farm director, about an innovative farming practice that he had helped develop. It was part of a new book I had decided to write about climate and carbon (published in 2014 as *Grass, Soil, Hope*). After the interview, I planned to head deeper into New England—*terra incognita* for this westerner. Despite having been born in Philadelphia and raised on a farm nearby for the first six years of my life (though not by farmers), I knew virtually nothing about this corner of America—a deficiency I hoped to correct.

I called ahead to Rodale and learned that my designated greeter had been hit in the head by a piece of airborne lawn furniture during the storm. I already knew that Jeff Moyer had gone home to start the generator at his farm, which, like thousands of other homes and businesses, had become temporarily detached from the nation's power grid. So, after depositing my luggage in the guesthouse, I headed into town for a meal and a beer. That's when the idea of what makes us come alive popped into my head. It was the beer; not the liquid substance itself—rather the *idea* of beer.

I'll back up.

While attending a conference the previous May, I heard a presentation by Katie Wallace, a young woman who worked for New Belgium Brewing Company in Colorado. Her official job title is sustainability specialist—no doubt a rapidly expanding career path for young people in the Age of Consequences—which means she gets paid to help the brewery increase its recycling program, reduce its carbon footprint, expand its use of sustainably raised hops and barley, analyze its water use, and generally make beer more earth friendly. It's not a marketing ploy, she insisted. The company is dedicated to core principles, including fair wages, good working conditions, environmental stewardship, and sustainability. She admitted, however, that when she took the job, she asked herself a question: What does "sustainability" actually mean? Doing research, she came up with a variety of conventional definitions, none of them particularly inspiring. She decided to reread New Belgium's published principles, discovering this instruction: have fun. That's when a lightbulb went off, she told the audience. Up came the next slide.

"It isn't sustainable if it isn't fun," she said.

That's when the lightbulb went off for me. She's exactly right. We can talk all day about "shoulds" and "musts" and "oughts" in relation to saving the planet, but in the end, if we're not having fun while doing these things, we won't succeed. It doesn't have to be deliriously, Pollyanna-ish fun necessarily, but it does need to be upbeat and enjoyable. People will respond much more quickly to a new idea if makes them smile. Conversely, if it's seen as a chore, no matter how well-intentioned, or as an obligation (especially if someone else is doing the obligating), then it will likely never be fully embraced. Then there's the burnout factor. If what you're doing isn't fun, *you* won't be sustainable for very long. That's a condition I knew all too well.

Of course, it's a lot easier to have fun when there's beer involved.

Ms. Wallace was also the source of the Howard Thurber quote, which I turned over in my mind as I drove into Kutztown. What makes us come alive, other than hurricanes and emergencies? What motivates humans to go do good things? We know what motivates us to do *bad* things—greed, for example—but how can we inspire people to come alive with joy and hope—other than drinking a lot of beer? And while I was thinking about it, what makes *me* come alive? Exploring did, certainly. For over fourteen years, I had explored a mysterious and fascinating land called Quivira, and I did so with gusto. It made me come alive. I was exploring new territory again, but it hadn't fired me up, not the way Quivira did. Maybe chasing Irene up the coast would help. Maybe among the flattened cornstalks and the broken branches left behind by the partying giants I could find what makes us come alive.

▪

FOOD MAKES US come alive.

I discovered this truth at the international slow food/Terra Madre gathering in Turin, Italy, a few years before, and I saw it again at Rodale, whose motto is, "Pioneering Organic Since 1947." The institute sits on 333 acres of a former industrially managed farm that is now dedicated to a holistic worldview based on *health*—soil, people, and communities. On a Tuesday morning, I was scooped up by Jeff and driven to a government office near Harrisburg to attend a day-long Organic Farming 101 workshop, where Jeff was one of the instructors. It was a great class. I learned about recycling nutrients, encouraging natural predators to manage pests, and increasing plant densities to block weeds—all integrated and interconnected. When livestock and poultry are added, the potentials rise even higher, due to fertilizing, grazing behavior, and culling unwanted plants. A lot of what I heard was technical, to be sure, but at its core was this message: organic makes the land come alive.

The message was buttressed that evening as I walked through Rodale's small on-site organic farm, which is open to the public. A mother and her two children were gathering fruit in the "U-Pick-It" section of the farm, and I could tell even from a distance that they were smiling. The kids frolicked while their mother filled a basket with ripe fruit—a timeless image made more timeless by the lovely light of the setting sun. We spoke briefly—I asked if I could take a photo of a child—and we concurred that fresh food is a good thing. Organic, and being outside.

This feeling was reinforced the next morning when Jeff and I discussed the no-till crimper that he had helped to develop at Rodale. I won't go into the details here. I'll just say this: five thousand years of plowing was a huge mistake. By turning soil over and exposing microbes to the killing effects of too much heat and light, plowing destroys life in the soil—which is contrary to everything that organic stands for as a worldview. That's why many farmers now employ a no-till drill to plant their seeds. However, many of them also use pesticides and herbicides to control bugs, weeds, and diseases. It was an unhappy paradox until Jeff came up with a way to do no-till organically, using a large rolling metal crimper to knock down the cover crop, keeping the weeds, bugs, and diseases way down. It was a brilliant innovation, and my hunch is that it will revolutionize agriculture someday. That's because by keeping the soil intact, it made microbes come alive.

The world needs soil microbes, and lots of them. According to the United Nations, there will be nine billion people on the planet by 2050, which raises a serious question: how are we going to feed them without destroying what's left of the natural world, especially under the stress of climate change? It's not about poor people and starvation either. The food well-fed Americans eat comes from a global production system that is already struggling to find enough arable land, adequate supplies of water, and drought-tolerant plants and animals to feed seven billion people. Add two billion more—of all income levels—and you have a recipe for a devastating raid on the natural world. Where is all this extra food and water going to come from, especially if the climate gets hotter and drier in many places as predicted? Organic, no-till agriculture is one answer. Healthy soil microbes is another. So too are innovators like Jeff Moyer and the Rodale Institute.

Healthy food not only makes us come alive, it *keeps* us alive as well.

▪

LIBERTY MAKES US come alive.

Giving Boston a wide berth while driving north to New Hampshire to visit another farmer, I caught sight of a sign for the Minute Man National Historical Park. I veered off the interstate and was immediately enmeshed in a thick tangle of SUVs, sedans, and trucks that clogged every road. Where were they all coming from? Where were they going? At one point, I needed to make a left-hand turn and waited patiently for a gap in the steady stream of traffic—but one never appeared. Amazed, I gave up and pushed on. Was it like this every day? Where did all these people live? How could they tolerate this grind-and-go existence? And why didn't they hit their brakes for a moment and let me make a left-hand turn? If this was the price of liberty, I wasn't sure it was worth it, I thought crankily.

I had freedom on my mind because I was trying to wend my way to the famous North Bridge near Concord, where on April 19, 1775, an assembly of American patriots, called minutemen, fired the "shot heard 'round the world" at a column of British soldiers, formally igniting the American Revolution. A quick detour to the park's visitor center set the stage: the British Army had occupied Boston in an attempt to suppress a rising tide of rebellion among American colonists. Receiving a tip that minutemen were caching weapons near Concord, the British decided to act. Marching under the cover of darkness, they hoped to catch the Americans by surprise and confiscate the weapons. Unfortunately for them, a silversmith named Paul Revere made a daring midnight ride to alert the militia, spoiling the surprise. Roused, the Americans gathered at the bridge for a confrontation with the hated British. Soon, shots rang out. Perhaps bloodshed was inevitable; in any case, the colonists' great struggle for liberty was officially underway.

Leaving the visitor center, I spotted a gap in the traffic, dashed into it, and headed for Concord. The weather was perfect. I stuck an elbow casually out the window as I drove. The park had a great story to tell, but it was difficult to cast my mind back two centuries amidst the high-priced homes, antique stores, and endless SUVs. I considered a stop in Concord, to soak up some colonial ambience, but I couldn't find a parking spot. I pushed on to North Bridge, a short distance to the west, where, at last, I found a bit of serenity. I climbed out of the rental and walked the short distance to the famous bridge, which earned its historical immortality mostly by being at the right place at the right time. It's a pretty spot, and as I mingled with two dozen other tourists, I tried once more to cast my mind back to those revolutionary days, when the quest for independence, freedom, and liberty made *a lot* of people come alive.

It was hard. It wasn't just the peaceful trees swaying in a soft breeze near the bridge, or the pleasant murmur of the creek as it passed underneath, that interfered with my attempt to imagine the battlefield. It was the general feeling that the revolution has stalled badly in the twenty-first century. American democracy is on the ropes, pummeled by the corrupting influence of big money, rancorous partisanship, poor leadership, widespread cynicism, and stubborn indifference on the part of the public. What would the minutemen have made of our political system today? They'd be baffled, I bet. Maybe angry too. Most distressing is our inability to get meaningful reform through Congress, other than health care (controversially too). That's because Business as Usual rules. Wasn't tyranny the Status Quo in 1775? Didn't Jefferson, Adams, Franklin, and fellow conspirators plot the ultimate reform by tossing the British out? Wasn't monarchy reformed with democracy? And wasn't the key to democracy the average citizen—the kind of guys who pointed their muskets at the hated Redcoats on the far side of North Bridge?

While eating supper in Providence, Rhode Island, later that evening, I read these words in the official National Park Service Handbook on the American Revolution:

> Our historical sites are so focused on the War for Independence that they give the visitors little sense of the wave of reform that swept America even while the battle against England wore on . . . Some of the most advanced reformist ideas came from the lower strata of American society. On farms, seaport docks, in taverns, and on streets, ordinary Americans were not only indispensable to the success of any reform movement but in many cases were the cutting edge of reform ideas.

Of all the reforms enacted at the time, the handbook went on to say, two were crucial: (1) enabling ordinary men to become active political players, both as voters and officeholders, and (2) severing the right to vote from property ownership. These reforms opened the floodgates of democracy, and both were quite unpopular with the upper classes of American society, as you can imagine. "We the people" began to think of themselves as the primary source of authority in the new nation instead of the elites, as well as the ultimate source of the nation's laws via their elected representatives. It was certainly something new under the sun, and the average American citizen at the time knew it. "Poor people," one eyewitness wrote, "enjoy the right of voting for representatives to be protectors of their lives, personal liberty, and their little property, which, though small, is yet, upon the whole, a very great object to them."

The revolutionary struggle for liberty led eventually to a host of other important reforms, including ending the policy of imprisonment for debt, the phasing out of indentured servitude, the end of slavery (requiring a terrible war, alas), the creation of taxpayer-funded public schools, the enactment of women's suffrage, the implementation of fair wages and working conditions for the laboring classes, the end of the poll tax, a loosening of divorce laws, and much more.

Many of these reforms spread around the world, of course, improving countless lives. Constitutionality, government neutrality in religion, the idea of inalienable rights of citizens, balances of power among governmental branches, limits to power, and other essentials all flowed from the cloud of gunpowder discharged from both ends of North Bridge on that fateful day. Democracy had come alive.

Was it still alive?

It's something to ponder as we move deeper into the Age of Consequences, with its unavoidable conflicts over natural resource scarcity combined with the social and ecological stress caused by climate change. Undoubtedly, the issues of liberty, inalienable rights, fairness, the pursuit of happiness, and perhaps more struggles with tyranny, will move to the front burner once more.

In many places around the world, they already have.

■

JUSTICE MAKES US come alive.

The distance between the American Revolution and the American Industrial Revolution is short, both historically and geographically—just sixty years and fifty miles. I'm referring to the distance between Concord and Lowell, Massachusetts, home to Lowell National Historical Park, where I made another quick stop. The park was created in 1978 to commemorate America's role in the Industrial Revolution, which by the 1820s was roaring ashore from England. What I didn't know was that Lowell was the first large-scale planned industrial city in American history. According to a brochure I picked up in the visitor center, in addition to being a "model city" and a major textile manufacturing center, Lowell ushered in a new era of mechanical innovation, gave rise to the modern corporation, and became a model (for a while) for America's new urban working class. It was also hailed as the Venice of the United States for its extensive canal system, used to power the mills.

I also learned about Lowell's prominent role in breaking the promise of American capitalism.

In the early days, the city's ten milling corporations staffed their thirty-two factories with unmarried Yankee farm girls lured to Lowell with promises of high wages, good working conditions, and a strict moral atmosphere. The promises were mostly true, at least for a while. It was a type of working-class Eden, where hard work, efficiency, and mechanization combined wholesomely to make the owners of the mills rich. Even Charles Dickens was impressed. In 1842, the famous novelist visited a factory at Lowell and reacted favorably. "I cannot recall or separate one young face that gave me a painful impression," he wrote later. "Not one young girl whom, assuming it to be matter of necessity that she should gain her daily bread by the labour of her hands, I would have removed from those works if I had had the power."

Apparently, he should have looked harder. Conditions in the mills were tough. Girls worked twelve-hour days and averaged seventy-three hours a week—all standing or sitting at machines that never ceased operating. The noise in the workrooms was described as "infernal," and the air was full of fibrous particles. The girls worked year-round with very little time off, though brief vacations were possible if a group pooled their labor and met their quotas. Often, the girls sent their wages home to their families, sometimes to support a brother's higher education. The girls paid rent to the mill owners to live in nearby boarding houses, six to a room, where they took their meals. Behavior was strictly regulated, as was a curfew. "Immorality" was not tolerated. Attending church on Sunday was a requirement.

This "Eden" didn't last long.

In the early 1830s, mill owners in neighboring cities began to cut costs by hiring workers from the pool of newly arrived immigrants to America, who worked (and lived) cheaply. This put pressure on the Lowell corporations to cut costs as well or have their products undersold. A critical decision point arose: cut worker wages or investor dividends? You know what happened. In 1834, the owners of Lowell's textile mills imposed a 15 percent reduction in wages. The Yankee farm girls promptly organized a "turn-out," or strike. This was a surprise to many observers, since it was considered at the time "unfeminine" to create such a fuss. In any case, the strike failed miserably, and within days, many of the girls were back at work at reduced pay. The rest went home, disillusioned. When the mill owners raised the rents at the boarding houses two years later, the girls "turned out" again, with greater effect. This time, the mill owners relented and rescinded the rent hike.

The handwriting was on the wall, however. Labor and capital would never again enjoy a harmonious relationship in this nation. Eden's jig was up. The Yankee girls kept up their protests, including a demand for a ten-hour day, which the mill owners resisted mightily. The fight spilled over in all directions. Soon, it became a historical struggle between the rights of workers and the power of corporations, with consequences that are still with us today. As the Yankee girls demonstrated, justice then, as now, is certainly worth coming alive for. Here are words from a Lowell protest song:

Oh! isn't it a pity, such a pretty girl as I
Should be sent to the factory to pine away and die?
Oh! I cannot be a slave, I will not be a slave,
For I'm so fond of liberty,
That I cannot be a slave.

I left the visitor center and took a quick walk around the grounds, pondering what I had just learned. Peeking around tall brick buildings and into narrow alleys, I searched for a clue that might reconcile the revolutionary fight for liberty fifty miles away at Concord with the homegrown oppression on exhibit at Lowell. I didn't find any. Lurking in the background, of course, is the twenty-first-century corporation, with all of its attendant power and oppression. A lot of people had come alive in recent years in this particular fight for justice—and I admired them all.

▪

CREATIVITY MAKES US come alive.

I didn't realize that Lowell was hometown to Jack Kerouac, the famous beat novelist of the 1950s and '60s. Of the more than thirty books of prose and poetry that he composed over his forty-seven-year life, five drew on his youthful adventures in Lowell and the French-Canadian working-class community he grew up in. I hadn't read any of them, I have to admit. I'm not a Kerouac fan, at least not of *On the Road* and *Dharma Bums*, the only two books of his that I've read. I thought they were indulgent and prosaic, recalling Truman Capote's famously malicious insult of Kerouac's work: "That's not writing, that's *typing*."

Wandering through the Kerouac display in the visitor center, however, I saw a different side of the writer. I saw a man possessed by his muse. He wrote and wrote and wrote. Words gushed out of him all his life, in books, journals, letters, and poems. He couldn't help himself, no more, apparently, than he could stop consuming alcohol. He wasn't a calculating writer—like Capote—or overtly fame seeking, like some of his contemporaries. He was a writer through and through. Words made him come alive—even if they also caused him to drink himself to death. His muse had him by his white T-shirt and wouldn't let go.

This thought stayed with me all the way to Portsmouth, New Hampshire, where I stopped for a break. It was another gorgeous day, so I took a seat at a table outside a small cafe, where, to my surprise, I heard the pleasing sounds of a violin among the noisy traffic. Looking around, I saw a young woman in a dress and cowboy boots not far away playing against the red brick wall of a fancy store, her case propped open for donations. I closed my eyes and listened. The music she played was as sweet as the soft air, carrying my spirits up and out someplace, to spin and float in the bright light. It cut through the traffic like, well, a song. Music makes us come alive, no doubt about it.

Over the centuries, I suspect, as many words have been written about the creative impulse in humans as have been composed about oppression and liberty. I won't repeat the arguments here, or add many words of my own, other than to say that music making and other creative endeavors is one of the very best ways humans come alive. In fact, it practically defines us as human. That's why the prehistoric cave art of Europe resonates so well with us today—it touches something very deep in our spiritual and psychological core. Ditto with music, dance, sculpture, and literature—as well as engineering, math, physics, and so on. We're an inventive species, obviously, and we've done some pretty spectacular things with our creative impulse, both good and bad. One of my favorite examples is Michelangelo, of Sistine Chapel and *David* fame, who epitomized the spirit of the Renaissance. What an extraordinarily alive person he must have been. But no less are the anonymous engineers behind the Apollo moon program, say. Imagine how alive they felt when Neil Armstrong kicked up a little lunar dust in July 1969.

All of this was very much on my mind over the next two days as I visited with Dorn Cox, a young self-described "carbon farmer" on his family's 250-acre farm, called Tuckaway, near Lee. I came to see Dorn because he was doing very innovative things, including organic no-till agriculture—all while earning a PhD. When I caught up with Dorn, he stood in a hayfield behind a University of New Hampshire professor's house, spreading wood ash carefully among a grid of study plots. His dissertation research aims at figuring out the best way to turn a hayfield into a vegetable farm without tilling it. Actually, he calls what he does "beyond organic no-till" because it tackles a variety of twenty-first-century challenges, including renewable energy. For example, Tuckaway produces 100 percent of its energy needs on-site, from only 10 percent of its land. Dorn does it with biodiesel—canola specifically, which he and his family grow on the farm. Additionally, Dorn's sister and her husband are avid practitioners of horse farming, like the Amish, which is another form of renewable energy.

Dorn also views much of what he does with an eye to climate change. By purposefully increasing the organic content of the farm's soil over time, Dorn and his family are sequestering additional carbon dioxide in the soil, thus helping to mitigate, in their small way, the carbon dioxide overload in our atmosphere. Improving the organic content of the soil is also good for their bottom line, because they can grow more crops. In fact, Tuckaway is involved in a Community-Supported Agriculture (CSA) program, coordinating its work with other farms nearby. Dorn is also deeply committed to open-source knowledge sharing via the Internet. Nothing's proprietary—all knowledge and experience is shared openly and evenly with anyone who wants access and vice versa. It's a democratic way, he told me, of making sure that everyone benefits.

Dorn's goal is nothing less than the revitalization of New Hampshire's moribund agriculture as a regenerative and sustainable enterprise, one acre at a time. Since the state currently produces only 6 percent of the food it consumes, and less energy, this dream is a tall order, to be sure. Dorn is undaunted, however. Tuckaway Farm is a passion for him, as well as a canvas of sorts for his ideas and creativity. He loves what he does, and clearly, what he is doing makes him come alive. It was all very exciting to see and hear, and it was hopeful in so many ways.

■

ALL GOOD STUFF—but what made *me* come alive?

I hadn't settled on an answer yet, so I decided to keep following in Irene's wake by making a long drive to a family haunt on Deer Isle, in Maine's Penobscot Bay. For a brief moment in the mid-1960s, my parents owned a tiny patch of shoreline near the village of Deer Isle, which we visited occasionally from our home near Philadelphia. I had no idea how it came into our possession or why my parents picked

Deer Isle of all places since we had no family roots in Maine. Perhaps my father wanted to play woodsman. I remember him chopping trees down on the property. Maybe the land represented some sort of dream of theirs, perhaps the dream of a vacation home (the tree chopping suggests they had planned to stay in the East). I also remember my mother sitting ruefully on a rock near the water's edge, looking out to sea.

We only visited the property a few times that I can recall, and I doubt it meant much emotionally to either my father or mother; at least they never said so. But the place stuck in my memory for some reason, perhaps because it belonged to us during one of the few happy periods in my mother's life that I witnessed. I knew she struggled all her life to find something that made her come alive. During the 1950s, she seemed to have her pick. She traveled, wrote, photographed, attended graduate school, worked, and played, though I suspect her inability to settle on one activity indicated a restless, questing heart as much as any drive for a career. Parenting certainly didn't make her come alive, as my sisters and I can attest. Writing did, however, as did traveling, but she largely abandoned both by the time I became a teenager. She wore her disappointment on her sleeve too. I think that's why the little property on remote Deer Isle has stayed with me all these years—it belongs to a period between alive and not alive in my mother's life. Did she come alive when they visited the property? Possibly—but I'll probably never know for sure.

I thought I'd go take a look.

It was nearly a mistake. Forgoing the more direct route up the interstate, I opted instead for the scenic drive along Route 1, which I picked up at Brunswick after a quick stop for caffeine. The road looked intriguing on my map, bobbing and weaving among inlets, bays, and rivers. What the map didn't show was the bumper-to-bumper traffic *all the way*. It took me over six hours to drive the one hundred miles to the island. I did the final twenty-five in a blur, partly because I wanted to arrive before the sun sank below the horizon, for photographic purposes, and partly because my bladder was about to burst.

I didn't know where our little property was located exactly, but I came close enough. After prospecting among the stately homes and "No Trespassing" signs that blanketed the shoreline where I knew the land should be (garnering a glower from a homeowner in his well-appointed backyard), I headed to Stonington, the main village on the island. No new memories had come swimming back during my prowl, which wasn't a surprise. I did get to see unfamiliar trees and a pretty view of the sea, however, which added shapes and colors to my mental images of our time there. I wasn't disappointed, despite the long drive. I was hungry, though, so I headed into town for a meal.

I found a table on a restaurant patio overlooking the bay. I heaved myself into a chair, ordered coffee to gird myself for the long drive back to my hotel in Portland, and turned to watch the day slip into evening. It was extraordinarily pleasant out, still, warm, and peaceful, marred only by a long string of profanity uttered by a crew hand unloading a shrimping boat nearby. It was an amusing juxtaposition to the art galleries and upscale restaurants that surrounded me. When the swearing ended, I soaked up the silence, relaxing for the first time all trip, it felt like.

The lack of relaxation was my fault—I push and push and push myself all the time, such as driving six hours through insane traffic to have a cup of coffee by the sea. It was part of the questing gene that I inherited from my mother, I suppose, along with the need to write. My thoughts drifted back to Mr. Kerouac and the restlessness that defined a generation—my mother's generation. They were only a few years apart in age, and I think they shared similar dreams and frustrations, and perhaps demons. One wrote, one wanted to. Both were restless, and both came alive for a while under the gaze of the same uncompromising muse, who I'll call . . . *Irene.*

The particular type of rootless, restless questing that came alive in the 1950s and '60s, fueled by various cultural disharmonies and cross-currents that Kerouac sensed early on, and my mother experienced first hand, is alive and well today. That's because the disharmonies have only grown bigger and more consequential, which means the questing is more necessary than ever. I know the feeling—I've been on a restless quest for most of my life. The actual object of the chase wasn't always clear to me, except it always involved asking questions and seeking answers. And expressing them as creatively as possible. It's what makes me come alive. The quest has one rule: obey the muse. Okay, a second: don't forget to have fun.

Preferably without the alcoholism.

In the Age of Consequences, we need as much creativity as we can muster. We'll need it not only to handle emergencies, produce fresh food, ensure liberty, and fight for justice, but because we'll need music to keep our spirits up and storytelling to communicate it for others to read and hear. As for me, hurricanes and other emergencies aren't my thing. I wasn't destined to be a farmer or rancher either, despite my huge admiration for everyone who grows our food and takes care of our land. Liberty and justice? Yes—but I'm not made of the sterner stuff required of activists. As I discovered, I'm not genetically predisposed for repeated tilting at Windmills. Writing is in my genes, however—and not just from my mother. In the 1980s, thanks to an aunt's sleuthing, I learned that William Faulkner is a not-too-distant cousin of mine on my father's side (to my mother's chagrin). I confirmed the relationship during a visit to Faulkner's home near Oxford, Mississippi, a few years ago. This knowledge has been a secret source of inspiration for me. Sitting there, smelling the salty air of the harbor, I decided to embrace my roots.

I raised my coffee cup in a toast. "To Irene, the muse of restless writers."

I gazed across the darkened bay. Lights twinkled in the distance. A crescent moon hung suspended over the shrouded outline of an island. It was time to hit the road. I finished the coffee and took one last look around. There's one more thing that makes us come alive, I thought: horizons. The need to see around the next bend. To hit the road. To chase a dream. And to never stop.

Irene would understand.

PART TWO

HOPE

FIFTEEN

THE NEW RANCH

(2002)

"Ranching is one of the few western occupations that have been renewable and have produced a continuing way of life."

—WALLACE STEGNER

IT WAS A bad year to be a blade of grass.

In 2002, the winter snows were late and meager, part of an emerging period of drought, experts said. Then May and June exploded into flame. Catastrophic crown fires scorched over a million acres of evergreens in the "four corner" states—New Mexico, Arizona, Colorado, and Utah—making it a bad year to be a tree too.

The monsoon rains then failed to arrive in July, and by mid-August, hope for a "green-up" had vanished. The land looked tired, shriveled, and beat-up. It was hard to tell which plants were alive, dormant, or stunned, and which were dead. One range professional speculated that perhaps as much as 60 percent of the native bunch grasses in New Mexico would die. It was bad news for the ranchers he knew and cared about, insult added to injury in an industry already beset by one seemingly intractable challenge after another.

For some, it was the final blow. Ranching in the American West, much like the grass on which it depended that year, has been struggling for survival. Persistently poor economics, tenacious opponents, shifting values in public-land use, changing demographics, decreased political influence, and the temptation of rapidly rising private land values have all combined to push ranching right to the edge. And not just ranching; according to one analysis, the number of natural-resource jobs, including agriculture, as a *share* of total employment in the Rocky Mountain West has declined by two-thirds since the mid-1970s. Today, less than one in thirty jobs in the region is in logging, mining, or agriculture. This fits a national trend. In 1993, the U.S. Census dropped its long-standing survey of farm residents. The farm population across the nation had dwindled from 40 percent of households in 1900 to a statistically insignificant 2 percent by 1990. The bureau decided that a survey was no longer relevant.

If the experts are correct—that the current multiyear drought could rival the decade-long "megadrought" of the 1950s for ecological, and thus economic, devastation—the tenuous grip of ranchers on the future will be loosened further, perhaps permanently. The ubiquitous "last cowboys," mythologized in a seemingly endless stream of tabletop photography books, could ride into their final sunset once and for all.

Or would they?

After all, for millions of years, grass has always managed to return and flourish. James Ingalls, U.S. Senator from Kansas (1873-1891) once wrote:

> Grass is the forgiveness of nature—her constant benediction. Fields trampled with battle, saturated with blood, torn with the ruts of cannon, grow green again with grass, and carnage is forgotten. Streets abandoned by traffic become grass grown like rural lanes, and are obliterated; forests decay, harvests perish, flowers vanish, but grass is immortal.

Few understand these words better than ranchers, who, because their cattle require grass, depend on the forgiveness of nature for a livelihood while simultaneously nurturing its beneficence. And like grass, ranching's adaptive response to adversity over the years has been patience—to outlast its troubles. The key to survival for both has been endurance—the ability to hold things together until the next rainstorm. Evolution favors grit.

Or at least it used to.

Today, grit may still rule for grass, but for ranchers, it has become more hindrance than help. "Ranching selects for stubbornness," a friend of mine likes to say. While admiring ranching and ranchers, he does not intend his quip to be taken as a tribute. What he means is this: stubbornness is not adaptive when it means rejecting new ideas or not adjusting to evolving values in a rapidly changing world.

This is where ranching and grass part ways ultimately—unlike grass, ranching may not be immortal.

Fortunately, a growing number of ranchers understand this and are embracing a cluster of new ideas and methods, often with the happy result of increased profits, restored land health, and repaired relationships with others. I call their work "the New Ranch"—a term I coined years back in a presumptuous attempt to describe a progressive ranching movement emerging in the region.

But what did it mean exactly? What were the new things ranchers were doing to stay in business while neighboring enterprises went under? How did they differ from new ranch to new ranch? What were the commonalities? What was the key? Technology, ideas, economics, increased attention to ecology, or all of the above?

During that summer of fire and heat, I decided to take a fourteen-hundred-mile drive from Santa Fe to Lander, Wyoming, and back, to see the New Ranch up close. I visited four families and was so inspired by what I saw and learned that I kept driving, in a sense, upon my return home. I needed to keep looking, listening, and learning. Since that summer, I have visited more ranchers, as well as environmentalists, scientists, and others, and asked more questions, all in a continuous quest for pieces to a jigsaw puzzle that eventually grew bigger than the New Ranch.

Initially, however, I wanted to know if ranching would survive this latest turn of the evolutionary wheel. Was it still renewable, as Stegner once observed, or were we destined to redefine a ranch as a mobile home park and a subdivision? But I also wanted to discover the outline of the future, and, with a little luck, find my real objective—hope—which, like grass, is sometimes required to lie quietly, waiting for rain.

THE JAMES RANCH
NORTH OF DURANGO, COLORADO

ONE OF THE first things you notice about the James Ranch is how busy the water is. Everywhere you turn, there is water flowing, filling, spilling, irrigating, laughing. Whether it is the big, fast-flowing community ditch, the noisy network of smaller irrigation ditches, the deliberate spill of water on pasture, the refreshing fish ponds, or the low roar of the muscular Animas River, take a walk in any direction on the ranch during the summer and you are destined to intercept water at work. It is purposeful water too, growing trees, cooling chickens, quenching cattle, raising vegetables, and, above all, sustaining grass.

All this energy on one ranch is no coincidence—busy water is a good metaphor for the James family. The purposefulness starts at the top. Tall, handsome, and quick to smile, David James grew up in Southern California, where his father lived the American Dream as a successful engineer and inventor, dabbling a bit in ranching and agriculture on the side. David attended the University of Redlands in the late 1950s, where he majored in business, but cattle got into his blood, and he spent every summer on a ranch. David met Kay, a city girl, at Redlands, and after getting hitched, they decided to pursue their dream: to raise a large family in a rural setting.

In 1961, they bought a small ranch on the Animas River, twelve miles north of the sleepy town of Durango, located in a picturesque valley in mountainous southwestern Colorado, and got busy raising five children and hundreds of cows. Durango was in transition at the time from a mining and agricultural center to what it is today: a mecca for tourists, environmentalists, outdoor enthusiasts, students, retirees, and real estate brokers. Land along the river was productive for cattle and still relatively cheap in 1961, though a new type of crop—subdivisions—would be planted soon enough.

Not long after arriving, David secured a permit from the United States Forest Service to graze cattle on the nearby national forest. The permit allowed him to run a certain number of cattle on a forest allotment. Once on the forest, he managed his animals in the manner to which he had been taught: uncontrolled, continuous grazing.

"In the beginning, I ranched like everyone else," said David, referring to his management style, "which means I lost money."

David followed what is sometimes called the "Columbus school" of ranching: turn the cows out in May, and go discover them in October. It's a strategy that often leads to overgrazing, especially along creeks and rivers, where cattle like to linger. Plants, once bitten, need time to recover and grow before being bitten again. If they are bitten too frequently, especially in dry times, they can use up their root reserves and die—which is bad news for the cattle (not to mention the plant). Since ranchers often work on a razor-thin profit margin, it doesn't take too many months of drought and overgrazing before the bottom line begins to wither too.

Grass may be patient, but bankers are not.

Through the 1970s, David's ranchlands, and his business, were on a downward spiral.

When the Forest Service cut back his cattle numbers, as they invariably did in years of drought, the only option available to David was to run them on the home ranch, which meant running the risk of overgrazing their private land. Meanwhile, the costs of operating the ranch kept rising. It was a no-win bind typical of many ranches in the West.

"I thought the answer was to work harder," he recalled, "but that was exactly the wrong thing to do."

Slowly, David came to realize that he was depleting the land, and himself, to the point of no return. By 1978, things became so desperate that the family was forced to develop a sizeable portion of their property, visible from the highway today, as a residential subdivision called, ironically, "the Ranch." It was a painful moment in their lives.

"I never wanted to do that again," said David, "so I began to look for another way."

In 1990, David enrolled in a seminar taught by Kirk Gadzia, a certified instructor in what was then called Holistic Resource Management—a method of cattle management that emphasizes tight control over the timing, intensity, and frequency of cattle impact on the land, mimicking the behavior of wild herbivores, such as bison, so that both the land and the animals remain healthy. "Timing" means not only the time of year but how much time, measured in days rather than the standard unit of months, the cattle will spend in a particular paddock. "Intensity" means how many animals are in the herd for that period of time. "Frequency" means how long the land is rested before a herd returns.

All three elements are carefully mapped out on a chart, which is why this strategy of ranching is often called "planned grazing." The movement of the cattle herd from one paddock or pasture to another is carefully designed, often with the needs of wildlife in mind. Paddocks can range from a few acres in size to hundreds of acres, depending on many variables, and are often created with permanent two-strand solar-powered electric fencing, which is lightweight, cost-effective, and easy on wildlife. It works too. Once zapped, cattle usually don't go near an electric fence again (ditto with elephants in Africa, as I understand it). Alternative methods of control include herding by a human (an ancient activity) and single-strand electric polywire, which is temporary and highly mobile. In all cases, the goal is the same: to control the timing, intensity, and frequency of the animal impact on the land.

Planned grazing has other names—timed grazing, management-intensive grazing, rapid rotational grazing, short-duration grazing, pulse grazing, cell grazing, or the "Savory system"—named after the Rhodesian biologist who came up with the basic idea.

Observing the migratory behavior of wild grazers in Africa, Allan Savory noticed that nature, often in the form of predators, kept herbivores on the move, which gives plants time to recover from the pressure of grazing. He also noticed that because herbivores tended to travel in large herds, their hooves had a significant ground-disturbing impact (think of what a patch of prairie would have looked like after a million-head herd of bison moved through), which he observed to be good for seed germination, among other things. In other words, plants can tolerate heavy grazing and perhaps even require it in certain circumstances. The key, of course, was that the animals moved on—and didn't return for the rest of the year.

Savory also observed that too much rest was as bad for the land as too much grazing—meaning that plants can choke themselves with abundance in the absence of herbivory and fire, prohibiting juvenile plants from getting established (not mowing your lawn all summer is a crude, but apt, analogy). In dry climates, one of the chief ways old and dead grass gets recycled is through the stomachs of grazers, such as deer, antelope, bison, sheep, grasshoppers, or cattle. Animals, of course, return nutrients to the soil in the form of waste products. Fire is another way to recycle grass, though this can be risky business in a drought. If you've burned up all the grass, exposing the soil, and the rains don't arrive on time—you and the land could be in trouble.

The bottom line of Savory's thinking is this: animals should be managed in a manner consistent with nature's model of herbivory.

David and Kay James did precisely that—they adopted a planned grazing system for both their private and public land operations. And they have thrived ecologically and economically as a result. They saved the ranch too—and today the four-hundred-acre James Ranch is noteworthy not only for its lush grass and busy water, but for its bucolic landscape in a valley that is dominated by development.

David and Kay insist, however, that adopting a new grazing system was only part of the equation, even if it had positive benefits for their bank account. The hardest part was setting an appropriate goal for their business. This was something new to the Jameses. As David noted wryly: "We really didn't have a goal in the early days, other than not going broke."

To remedy this, the entire James clan sat down in the early 1990s and composed a goal statement. It reads:

> The integrity and distinction of the James Ranch is to be preserved for future generations by developing financially viable agricultural and related enterprises that sustain a profitable livelihood for the families directly involved while improving the land and encouraging the use of all resources, natural and human, to their highest and best potential.

It worked. Today, David profitably runs cattle on 220,000 acres of public land across two states. He is the largest permittee on the San Juan National Forest land, north and west of town. Using the diversity of the country to his advantage, David grazes his cattle in the low (dry) country only during the dormant (winter) season; then he moves them to the forests before finishing the cycle on the irrigated pastures of the home ranch.

That's enough to keep anybody incredibly busy, of course, but David complicates the job by managing the whole operation according to planned grazing principles. Maps and charts cover a wall in their house. But David doesn't see it as more work. "What's harder," he asked rhetorically, "spending all day on horseback looking for cattle scattered all over the county, like we used to, or knowing exactly where the herd is every day and moving them simply by opening a gate?"

It's all about attitude, David observed. "It isn't just about cattle," he said, "it's about the land. I feel like I've finally become the good steward that I kept telling everybody I was."

Recently, the family refined their vision for the land and community one hundred years into the future. It looks like this:

- "lands that are covered with biologically diverse vegetation"
- "lands that boast functioning water, mineral, and solar cycles"
- "abundant and diverse wildlife"
- "a community benefiting from locally grown, healthy food"
- "a community aware of the importance of agriculture to the environment"
- "open space for family and community"

And they have summarized the lessons they have learned over the past dozen years:

- "Imitating nature is healthy."
- "People like to know the source of their food."
- "Ranching with nature is socially responsible."
- "Ranching with nature gives the rancher sustainability."

But it wasn't all vision. It was practical economics too. For example, years ago, David and Kay told their kids that in order to return home, each had to bring a business with him or her. Today, son Danny owns and manages a successful artisanal dairy operation producing fancy cheeses on the home ranch that he began from scratch; son Justin owns a profitable BBQ restaurant in Durango; daughter Julie and her husband John own a successful tree farm on the home place; and daughter Jennifer and her husband grow and sell organic vegetables next door and plan to open a guest lodge across the highway.

In an era when more and more farm and ranch kids are leaving home, not to return, what the James clan has accomplished is significant. Not only are the kids staying close; they are also diversifying the ranch into sustainable businesses. Their attention is focused on the modern West, represented by Durango's booming affluence and dependence on tourism. Whether it is artisan cheese, organic produce, decorative trees for landscaping, or a lodge for paying guests, the next generation of Jameses has their eyes firmly on new opportunities.

This raised a question. The Jameses enjoy what David calls many "unfair advantages" on the ranch—abundant grass, plentiful water, a busy highway right outside their front door, and close proximity to Durango—all of which contribute to their success. By contrast, many ranch families do not enjoy such advantages, which made me wonder: Beyond its fortunate circumstances, what can the James gang teach us?

I posed the question to David and Kay one evening.

"The key is community," said Kay. "Sure, we've been blessed by a strong family and a special place, but our focus has always been on the larger community. We're constantly asking ourselves, 'What can we do to help?'"

Answering their own question, David and Kay James decided ten years ago to get into the business of producing and selling grass-fed beef from their ranch—to make money, of course, but also as a way of contributing to the quality of their community's life.

Grass-fed, or "grass-finished," as they call it, is meat from animals that have eaten nothing but grass from birth to death. This is a radical idea because nearly all cattle in America end their days being fattened on corn (and assorted agricultural byproducts) in a feedlot before being slaughtered. Corn enables cattle to put on weight more quickly, thus increasing profits, while also adding more "marbling" to the meat—creating a taste that Americans have come to associate with quality beef. The trouble is that cows are not designed by nature to eat corn, so they require a cornucopia of drugs to maintain their health.

There's another reason for going into the grass-fed business: it is more consistently profitable than regular beef. That's because ranchers can market their beef directly to local customers, thus commanding premium prices in health-conscious towns such as Durango. It also provides a direct link between the consumer and the producer—a link that puts a human face on eating and agriculture.

For David and Kay, this link is crucial—it builds the bonds of community that hold everything together. "When local people are supporting local agriculture," said David, "you know you're doing something right."

Every landscape is unique, and every ranch is different, so drawing lessons is a tricky business, but one overarching lesson of the James Ranch seems clear: traditions can be strengthened by a willingness to try new ideas. Later, while thumbing through a stack of information

David and Kay had given me, I found a quote that seemed to sum up not only their philosophy, but also that of the New Ranch movement in general and the optimism it embodies. It came from a wall in an old church in Essex, England:

A vision without a task
Is but a dream.
A task without a vision
Is drudgery.
A vision and a task
Is the hope of the world.

THE ALLEN RANCH
SOUTH OF HOTCHKISS, COLORADO

STAND ON THE back porch of Steve and Rachel Allen's home on the western edge of Fruitland Mesa, located 150 miles north of the James Ranch, in the center of Colorado's western slope, and you will be rewarded with a view of Stegnerian proportions: Grand Mesa and the Hotchkiss Valley on the left, the rugged summits of the Ragged Mountains in the center, and on the right, the purple lofts of the West Elks, a federally designated wilderness where Steve conducts his day job. Like the James Ranch, the Allens are permittees on the national forest, but what made them unique was *how* they grazed on public land—they herded their cattle

I met Steve three years earlier at a livestock-herding workshop I organized at Ghost Ranch, in northern New Mexico. I knew that his grazing association, called the West Elk Pool, had recently won a nationwide award from the Forest Service for its innovative management of cattle in the West Elk Mountains. The local Forest Service range conservationist, Dave Bradford, had won a similar award for his role in the West Elk experiment. Intrigued, I invited them both down to speak about their success.

Steve began their presentation that day with a story. Driving to the workshop, he said, he and Dave found themselves stuck behind a slow-moving truck on a narrow, winding road. At first they waited calmly for a safe opportunity to pass, but none appeared. Then they grew impatient. Finally, they took a chance. Crossing double yellow lines, they hit the accelerator and prayed. They made it—luckily there had been nothing but open road ahead of them, he said. It was meant as a metaphor—describing Steve's experience as a rancher and Dave's experience with the Forest Service. The slow-moving obstacle, of course, was tradition.

In the mid-1990s, Steve and Dave convinced their respective peers to give herding a chance in the West Elks. They proposed that six ranchers on neighboring allotments, each of whom ran separate operations in the mountains, combine their individual cattle herds into one big herd and move them through the wilderness in a slow, one-way arc. By allowing cattle to behave like the roaming animals that they are (or used to be), Dave and Steve argued, the plants would be given enough time to grow before being bitten again, which in the case of the cattle of the West Elk Pool wouldn't be until the following summer.

There were other advantages, as they discovered. From the Forest Service's perspective, having one herd on the move in the West Elks rather than six relatively stationary herds was attractive for ecological and other reasons, including reduced conflicts with wildlife. For the ranchers, one big herd cut down on the costs of maintaining fences and watering troughs. It was also less labor-intensive, though it didn't seem that way initially. On the first go-around, Steve recounted, they had twenty people working the herd, which proved to be about twelve too many. Today, they move the herd with two to six people and a bevy of hardworking border collies.

So why was all this so unusual? It is customary practice for ranchers to spread their cattle out over a landscape, especially in times of drought, not bunch them up. It's the Columbus school again—less management is the norm, not more. And herding means more management, even if it requires less people, which might seem counterintuitive. Herding is different because it is less dependent on *things*—fences, troughs, and other infrastructure—and more dependent on *people.* Not only is it an ancient human activity (think Persian nomads), but it dominated the early days of the Old West as well (think *Lonesome Dove*). Herding faded away, however, with the arrival of the barbed-wire fence and, later, the federal allotment system for grazing permits, both of which splintered the wide-open West into discrete units that lent themselves to a less-intensive management style.

Dave and Steve had turned the clock back—or forward, depending how one looked at it. At the workshop, they described the pattern of the herd's movement as looking like a large flowing mass, with a head, a body, and a tail, in almost continuous motion. Pool riders don't push the whole herd at once; instead, they guide the head, or the cattle that like to lead, into areas that are scheduled for grazing. The body follows, leaving only the stragglers—those animals who always seem to like to stay in a pasture—to be pushed along.

The single-herd approach allows the ranchers to concentrate their energies on all of their cattle at once, as well as allowing the Forest Service to more easily monitor conditions on the ground.

In fact, the monitoring data showed such an improvement in the health of the land over time that the West Elk Pool asked for, and was granted, an increase in their permitted cattle numbers from the Forest Service. In other words, since the data supported their contention that herding was improving the health of land that had been beaten up historically by livestock, the ranchers of the West Elk Pool felt it was time to gain financially from their good work. This was significant because the trend in cattle numbers on public land was mostly the other direction—down—for a variety of reasons, not the least of which was pressure from environmentalists who saw cattle as simply destructive to land. But the West Elk Pool was different in this regard as well. In the planning process, they involved a local environmental group and ultimately got its blessing for the herding experiment.

It was a strategy that paid off, literally.

Once again, as with David and Kay James, it was about a vision. After the herding workshop, Dave sent me the goal statement for the West Elk allotment, which read in part:

> Our goal is to maintain a safe, secure rural community with economic, social, and biological diversity . . . that respects individual freedom and values education, and that encourages cooperation . . . Our goal is to have a good water cycle by having close plant spacing, a covered soil surface, and arable soils; have a fast mineral cycle using soil nutrients effectively; have an energy flow that maximizes the amount of sunlight converted to plant growth and values the seclusion and natural aesthetics of the area.

Standing on the Allens' back porch, I asked Steve the question that had been on my mind since the workshop: What set him up for crossing those double yellow lines? A slight, quiet, but affable man, Steve didn't strike me at first as the ringleader type. Spend time with him, however, especially as he gently but firmly works his beloved border collies—he is a well-known trainer in the area—and you get the sense that a strong will is at work. Still, what leads someone to step out of the box like that?

Steve grew up in Denver, he said, where his father was an insurance salesman. He met Rachel at Western State College in Gunnison, where they discovered that they both liked to ski—a lot. Steve joined the ski patrol in Crested Butte, and eventually both of them became ski instructors. It was 1968. They were young and living the easy life. But restlessness gnawed at Steve. "The ski industry is designed to make ski bums, not professionals," he said with his easy smile. "It was fun, but we wanted more."

They were also restless about the changes happening in Crested Butte. Even in those early days, signs of gentrification were visible in town. Although not yet affected by the scale of change that transformed nearby Aspen into a playground for the rich and famous—a process sometimes called the "Aspenization" of the rural West—Steve and Rachel could see the handwriting on Crested Butte's wall. By the early 1970s, they decided to join the back-to-the-land movement, trading their skis for farm overalls.

"We weren't hippies, mind you," interjected Rachel, laughing. "We took farming seriously. I just want to get that on record."

They moved west, over the West Elks, to the small village of Paonia, where they planted what eventually became a large garden. They grew vegetables, raised chickens, produced hay—and learned from their farm neighbors.

"Because we admitted we didn't know very much," said Steve, "and because we were willing to learn, people were willing to teach."

Which could be a motto for the New Ranch.

In 1977, restlessness struck again. They traded the garden for a rundown farm on the edge of Fruitland Mesa, where the hay was so bad the first few years they had to give it away. Eventually, they bought a few cattle and decided to try their hand at ranching. In 1988, Steve purchased a Forest Service permit in the nearby West Elks, mostly as a forage reserve for his animals in times of drought. His interest was not purely economic, however. Steve had always been attracted to mountains, and soon he had a chance to work in them daily.

Eager to learn more, Steve took a Holistic Resource Management course the same year that Dave James did, and that led him to give herding a try. With the arrival of Dave Bradford to the Forest Service office in Paonia a short while later, the opportunity to cross the yellow lines suddenly presented itself.

As part of the process of pulling the West Elk experiment together, Steve also became a student of a new method of low-stress livestock handling sometimes called the "Bud Williams school"—after its Canadian founder. Its principles fly in the face of traditional methods of cattle handling, which are full of whooping, prodding, pushing, and cursing. Putting stress on cattle is as customary to ranching as a lasso and spurs.

But that was Steve's point: customary, yes; natural, no. And that's where herding comes in: pressure from predators in the wild made grazers bunch in herds naturally. Unfortunately, on many ranches today, the herd instinct has been prodded out of most cattle.

The whole idea of low-stress handling is to use a "law of nature" to positive effect.

"Nature," Steve said simply, "has the right ideas, but we keep messing them up."

It is this return to nature's original model—such as grass-fed livestock and low-stress herding—that defines the progressive ranching movement underway today. Ranching needs good students, but it needs good teachers too. It needs people like Steve Allen. Grass may lie patiently for rain, but people need inspiration.

TWIN CREEK RANCH

SOUTH OF LANDER, WYOMING

A FEW HOURS north of Fruitland Mesa, I entered the dry heart of that summer's drought, which was centered on southern Wyoming. After crossing Sweetwater River on the old Mormon Trail, I knew the precise moment when I had reached Tony and Andrea Malmberg's ranch. Rounding a big bend in the road, I was suddenly confronted with the sight of green grass, tall willows, sedges, rushes, and flowing water.

I had arrived at Twin Creek. I was here not only because I knew the Malmbergs to be first-rate land stewards, but also because of their experiences in diversifying their business, which seemed to be a requirement for success as a rancher these days.

Following the lush creek on the road toward the ranch headquarters, I recalled an anecdote Tony had included in an article titled "Ranching

For Biodiversity" that he had recently written for the Quivira Coalition's newsletter. It detailed an experience from his youth when he and a brother-in-law decided to blow up a beaver dam on the creek:

> Jim and I crawled through the meadow grass under his pickup giggling. Jim pulled the wires in behind him, leading to the charge of dynamite.
>
> "This will show that little bastard," I said. Jim touched the two wires to the battery. *WOOMPH!* The concussion preceded the explosion. Sticks and mud came raining down on the pickup. As soon as it stopped hailing willows and mud, we scrambled out from under our shield.
>
> "Yeah!" I hollered as we ran down the creek bank. "I think we got it all."
>
> Water gushed through the gutted beaver dam, and we could see the level dropping quickly. The next morning, I rode my wrangle horse across the restored crossing. The water behind the beaver dam had gotten so deep I couldn't bring the horses across. But that was taken care of now. I galloped down the creek. The water ran muddy, and I couldn't help but notice creek banks caving into the stream.
>
> I wondered.

It was another story about tradition, this time about conventional attitudes toward wildlife. But it was also an allegory. Years later, when Tony was a young man, his family ranch "caved in" too—forced into bankruptcy by high interest payments on loans and tumbling cattle prices, costing Tony's family the entire thirty-three-thousand-acre property. Suddenly homeless, Tony began to wonder what had happened. Two years later, he leased the ranch back from the new owner, before eventually buying it. But he knew things had to be different this time if he wanted to stay.

Like David James and Steve Allen, Tony attended a course on Holistic Resource Management and he began to realize that biodiversity was a plus on his ranch, not a minus. "I shifted my thought process to live with the beaver and their dams," he wrote in his article. "With this commitment, I viewed the creek as a fence rather than something I could cross. This attitude gave me an extra pasture, a higher water table, less erosion, and more grass in the riparian area. The positive results energized me, and I began to curiously watch in a new way."

What he noticed as a result of his new land management was an increase in biodiversity. Moose, previously a rare sight on the property, began to appear in larger numbers. He even began to appreciate the coyotes and prairie dogs on the ranch and the role they played in the health of his land. Later, a University of Wyoming study found a 50 percent increase in bird populations over the span of a few years.

All of which led him to formulate two guiding principles:

> First, I avoid actively killing anything, and notice what is there. Whether a weed or an animal, it would not be here if its habitat were not. I plan the timing, intensity, and frequency of tools (grazing, rest, fire, animal impact, technology, and living organisms) to move community dynamics to a level of higher diversity and complexity.
>
> Second, I ask myself what is missing. Problems are not due to the presence of a species but rather the absence of a species. The absence of moose meant willows were missing, which meant beaver were missing, and the chain continues.
>
> If I honor my rule of not suppressing life, I will see beyond symptoms to address problems. If I continue asking, "What is missing?" I will continue to see beyond simple systems and realize the whole. When I increase biodiversity, I improve land health, I improve community relations, and I improve our ranch profitability.

To accomplish his goals, Tony employs livestock grazing as a land-management tool. To encourage the growth of willows along the stream and ponds, for example, he grazes them in early spring to assist seedling establishment. By concentrating cattle for short periods of time in an area, Tony breaks up topsoils and makes the land more receptive to natural reseeding and able to hold more water.

What brought me to Twin Creek, however, wasn't just the tall grass, the flowing water, or even the progressive ranch management practiced by Tony and Andrea, though these were important. What I wanted to see was the very nice bed-and-breakfast they operated.

As I pulled up to the spiffy new three-story lodge, I was greeted with a sunny wave by Andrea. A child of the Wyoming ranching establishment—her father traded cattle for a living—Andrea heard Tony speak passionately some years earlier about the benefits of planned grazing at a livestock meeting (where his talk was coolly received) and wrote him an equally passionate letter challenging his beliefs. They corresponded back and forth until she accepted his dare to come to the ranch and see the proof herself.

Tony joined us inside the sunny lodge. Bearded, deep chested, and sporting a leather vest, Tony looks the part of the cowboy. He is also cheery and garrulous, in print and in person.

Over a glass of wine later that evening, I learned that the lodge is the happy ending to a story that had its roots in anger. "When my family lost the ranch," recalled Tony, "I blamed everyone but ourselves. I blamed consumers, environmentalists, liberals. But most of all, I blamed our new neighbors."

In 1982, as the family was slipping into bankruptcy, a man from California bought a neighboring ranch for twice what a cow would generate per acre. Although this fact didn't directly affect his family's pending insolvency, it angered Tony because it suggested the end of an era. Ranch land had more value to society, he saw suddenly, as an amenity than as a working landscape. Recreation trumped ranching. And Tony didn't like it.

But then Tony had a revelation: markets don't lie. Upon returning to the ranch, he decided that in addition to the cattle operation, he would start a ranch-recreation business and market stays directly to people who wanted the cowboy experience. He quickly learned, however, that paying guests wouldn't tolerate dirt or mice as much as he did, so he and Andrea took the plunge and built a pretty lodge with a capacity for fourteen guests at a time.

But they didn't stop there. Making it economically meant exploring as many diverse business enterprises as possible. Andrea convinced Tony that the next step was to "go local" and find ways to tap local markets, including their new neighbors, for their beef and other services. They hosted a class on weed control for local ranchette owners and focused on the positive role of goats—which will eat every noxious weed on the state list. It was a big hit.

That was followed by a seminar on rangeland health, which proved popular with their ranching friends. Then came a foray into the grass-fed beef business, which has been successful too.

Next in their efforts at economic diversification was Andrea's decision to start teaching yoga. A recent winter solstice party packed the lodge with what Tony called the "strangest assortment of people I'd ever seen together." He continued, "The hodgepodge appeared to be a demographic accident, yet they all ended up in central Wyoming because they wanted the same things we want: a beautiful landscape, healthy ecology, wholesome food, and a sense of community." In this, Tony drew a parallel with the benefit of increased biodiversity on the ranch.

"In the old days, I didn't have to deal with people different from me," he said. "But this is better."

Tony went on to explain to me how his indicators of success have changed over the years. In 1982, his primary measure of success was a traditional one: increased weaning weights of his calves. By 1995, Tony's measure had shifted to the stocking rate of cattle (the number of cattle per acre on the ranch—more cattle, managed sustainably, equals greater profitability), which, thanks to planned grazing, was up 75 percent from years prior. By 1998, his indicator had shifted to monitoring—data produced by a detailed study of plants—and what it said about ecological trends on the ranch. In his case, the trend was up—which was a good sign. By 2000, Tony used the diversity of songbirds on the property as his baseline (over sixty species currently). By 2002, however, the main measure of success had changed to an economic one: how many activities generated income for the ranch in a year. At the time of my visit, they were up to three.

Tony attributes this success to their ability to speak different languages to different audiences, including recreationalists.

"I realized that if I'm going to survive in the twenty-first century, I need to be trilingual," Tony explained. "Ranchers tell stories. The BLM wants to talk data. And then we've got the environmentalists. Lander has a lot of them. To connect with them, you need to use poetry."

In other words, success in ranching today is as much about communication and marketing as it is about on-the-ground results. As Tony and Andrea's story suggests, it is not enough simply to *do* a better job environmentally, even if it brings profitability. One must also *sell* one's good work and do so aggressively in a social climate of rapid change and the general population's increasing detachment from our agricultural roots.

From all the indicators that I saw, Tony and Andrea are on the right track. The lodge was clean, comfortable, and airy; the food wonderful; and the visitors happy. But this is no dude ranch. Tony makes his guests work. According to his planned grazing schedule, his cattle need to be moved almost every day—so he has paying guests do it. They love it, of course, and since his cowboy does the supervisory work, Tony is free to explore other business ideas. And the ideas keep coming.

RED CANYON RANCH
WEST OF LANDER, WYOMING

WHEN I MET Bob Budd at the Nature Conservancy office in Lander, a short drive north from the Malmberg's Twin Creek Ranch, he was pacing the floor, waiting for my arrival.

"The ranch is on fire," he said quickly. "Let's go."

Despite being a foot taller than Bob, I had to hustle to keep up with him as we headed outside. A Wyoming native son, a member of a well-known ranching family, and former executive director of the state's cattlemen association, Bob managed the Red Canyon Ranch for the Nature Conservancy's Wyoming office when I met him. He also served as their director of science. Bob had earned a master's degree in ecology from the University of Wyoming and was in line to become president of the Society for Range Management, a highly respected national association of range professionals.

Without a doubt, he was a man on the move.

I jumped into my truck and followed Bob rapidly to the headquarters of the thirty-five-thousand-acre Red Canyon Ranch, which borders Lander on the south and west. The Nature Conservancy, Bob said, had purchased the property for three reasons: to protect open space and the biological resources held there, to demonstrate that livestock production and conservation are compatible, and to work at landscape-level management and restoration goals.

The first two goals have more or less been achieved, he said as I climbed into his truck in the parking lot. It is the third goal that motivated him now. What Bob wants is fire back on the land, brush and trees thinned, erosion repaired, noxious weeds eradicated, perennial streams to flow fuller, riparian vegetation to grow stronger, and wildlife populations to bloom.

And judging by the speed at which we traveled, he wanted them all at once.

Bob was thrilled about the lightning-sparked fire that was burning a chunk of forest and rangeland right where he had been encouraging the Forest Service to light a prescribed burn for years. That's because fire is a keystone ecological process, meaning a process that is fundamental to the health of the ecosystem over time. Research shows that "cool" fires happened frequently in Western forests, perhaps as often as every ten years in some stands. But for much of the twentieth century, humans suppressed all fires in our national forests, mostly to protect the monetary value of the timber, and as a consequence, the forests have become overgrown and dangerously prone to "very hot," destructive fires. To reverse this condition and restore forest health, ecologists and others have encouraged the Forest Service to light controlled burns. To many, however, the pace of bureaucracy has been frustratingly slow.

"I love lightning," Bob said with a twinkle in his light blue eyes, "because there's no paperwork."

As we sped into the mountains in search of a suitable vantage point to observe the progress of the fire, talking energetically about ecological theories that I had only recently begun to study, I recalled something Bob had recently written: "I am an advocate for wild creatures, rare plants, arrays of native vegetation, clean water, fish, stewardship of natural resources, and learning. I believe these things are compatible with ranching, sometimes lost without ranching. Some people call me a cowboy. A lot of good cowboys call me an environmentalist."

Bob has strong words for both camps, especially about their respective defense of myth. He likes to remind environmentalists in particular that nature is not as pristine as many assume. For thousands of years, he observed, Wyoming has been grazed, burned, rested, desiccated, and flooded. In saying so, he consciously tilts at an ecological holy grail called the "balance of nature." This is the long-standing theory that says nature tries hard to hold things in balance; in other words, when a system gets out of balance, nature works to right the ship, so to speak. Predator/prey populations are a good example. According to this theory, too many coyotes and not enough jackrabbits, say, mean nature will bring the coyote population back into balance over time (by starvation).

Today, most professional ecologists reject this theory in favor of one called the "flux of nature," which views nature as dynamic, chaotic, and rife with bouts of disturbance—such as forest fires and floods. Unfortunately, the "balance of nature" theory persists among nonprofessionals, especially environmentalists, resulting in a great deal of conflict with rural residents over ideas of proper stewardship.

"In landscapes where the single ecological truth is chaos and dynamic change," Bob wrote, "we seem obsessed with stability. Instead of relishing dynamic irregularities in nature, we absorb confusion and chaos into our own lives, then demand that natural systems be stable."

He likes to explain to both environmentalists and ranchers that grazing, like fire, is a keystone process. "Like fire, erosion, and drought, grazing is a natural process that can be stark and ugly," he wrote. "And, like fire, erosion, and drought, grazing is essential to the maintenance of many natural systems in the West . . . And because adults tend to overlook other grazing creatures, we forget the impact of grasshoppers, rodents, birds, and other organisms that have long shaped the West."

Just as prescribed fire, once controversial, is now widely accepted, Bob observed, it is simply a matter of time before the same change of thinking happens to grazing.

As we sped through the forest, still searching for a spot to view the fire, I asked him if he thought environmentalists would ever embrace ranching.

"I think they'll have to," he replied, "if they want to protect open space."

Bob explained that in Wyoming, like much of the West today, unbridled development on private land has resulted in habitat fragmentation and destruction. When land is subdivided, the new roads and homes often interrupt wildlife migration corridors, decrease habitat for rare plants and animals, and make ecosystem management difficult. The open space that ranches provide are the last barrier to development in many places. "The economic viability of ranching is essential," he said, "in maintaining Wyoming's open space, native species, and healthy ecosystems."

"Even on public lands?" I asked.

"Absolutely," he replied. "That's because it's all about proper stewardship. I don't care where you are."

Bob pointed to the trees outside the truck window.

"Our common goal must be to provide the full range of values and habitat types that a variety of species need, including us," he said. "And ranchers can help."

What he meant, I've come to understand, is that ranchers can become restorationists because they are uniquely positioned to deliver ecological services—food, fuel, fiber, and other ecological benefits that society requires—as landowners, as livestock specialists, and as hard workers. This will become increasingly important, I'm convinced, as the twenty-first century wears on and we come to realize just how much restoration work is required—not to restore the balance of nature but to get nature back into a position where it can operate according to natural principles, including disturbance. Cows can have a role here too. As domesticated animals, they can be used effectively to recreate certain kinds of animal impact on the land—a point Allan Savory made years ago.

Suddenly, we stopped. The fire we sought had proved elusive, and it was time to head back to headquarters. It seemed symbolic. While landscape-scale opportunities for ranchers may be plentiful, as Bob suggested, many are elusive, especially on public land, where every action seems to engender an opposite reaction by someone. Even the smallest restoration project, whether it involves livestock or not, can quickly become mired in red tape and conflict. Bob remained optimistic, however. He admitted that he had to be.

Returning to the ranch headquarters, Bob kept moving. He needed to take his son to baseball practice. I followed him into the house for introductions to the family. We talked for a while longer, shook hands, and before I knew it, he was gone.

Rather than drive off immediately too, I walked down to a bridge that spanned a burbling creek. Enjoying a momentary respite from the dust, the cascade of ideas, goals, and practices that dominated conversation for the entire trip, I leaned on the wooden railing and listened to the wind.

One thread that tied Bob Budd to the Jameses, Allens, and Malmbergs, it occurred to me, was the desire to make amends with nature. To paraphrase President John F. Kennedy, each asked not what the land could do for them, but what they could for the land. Whether it was restoring land to health, bridging urban–rural divides, teaching, feeding, or peacemaking, every person I encountered was engaged in an act of redemption, mostly by trying to heal damaged relationships, particularly our bond with the land. This is good news for grass, especially in these dry times. It is good news for all of us as well.

Grass may seem immortal, but in reality, it needs water, nutrients, animals, and fire to stay vigorous. The health of the whole depends on the health of its essential parts. This is important, as Bob Budd explained, because disruption is inevitable in nature; sooner or later, a calamity of some sort will strike, and those plant and animal populations that are not functioning properly at basic levels will be in jeopardy. Communities of people are no different. Whether it is a ranch, village, small town, or city, every community needs to be diverse, resilient, opportunistic, and self-reliant if it is to survive unexpected challenges.

For example, by setting water to work with a purpose—to earn a living within nature's model—the James family has buffered themselves well against uncertainty, and in the process protected four hundred acres of prime land along the Animas River from subdivision. The potential financial gain from busting their land into small lots for houses is astronomical—but they won't do it because it doesn't fit their goal for their family, their land, or their community.

Take Steve Allen, for example. It took the brave step of crossing yellow lines to achieve his goal. How many of us city folk are willing to take a risk like that? Do we even know where the yellow lines are? Are *we* resilient in our own lives? Or are we in spiritual (as well as practical) danger of supposing, as Aldo Leopold warned, that, "Breakfast comes from the grocery, and heat comes from a furnace."

Another thread was Tony Malmberg's question about the sanctity of life—when might we stop killing things we don't understand, as he did, and start inquiring instead about what might be missing from our lives? And once the outlines of answers become perceptible, what language do we speak so the lessons we've learned can be clearly understood? Must we be trilingual, or at some point will one vocabulary suffice—the language, say, of grass? Or food? And if we can figure all that out, how do we make it *pay*—as in paychecks—without which little can be accomplished.

Then there was the big picture. How do we work at scale, as Bob Budd advocates—and not just on ranches and farms, but all over the West, the nation, the globe? How do we take a landscape perspective in a world balkanized into countless and often feuding private, state, tribal, and federal fiefdoms? How do we overcome the paperwork, the lawsuits, the power struggles, and the politicking necessary to get the big work done in a century that will likely be roiled by climate change, energy instability, water shortages, and a host of other potential challenges?

I found the clues at nearly every stop along this trip. Stegner was right, ranching *is* renewable—in fact, it feels very much like it's being reborn, one ranch at a time. This is good news. Grass and grazers, after all, are the original solar power. Moreover, humans have been living and working with livestock for a very long time and through a great deal of historical change. The human desire to be near animals, and be outdoors, hasn't altered much over the centuries, though it has recently shrunk, hopefully temporarily, as a result of industrialization. We need ranching, I came away thinking, because it can be regenerative, not only for the food and good stewardship it can provide, but also for the lessons it can teach us about resilience and sustainability. All flesh is grass, as the Bible reminds us, though it has often been forgotten.

Perhaps it is time to consider it again.

SIXTEEN

AN INVITATION TO JOIN THE RADICAL CENTER

(2003)

FOR MORE THAN thirty years, environmentalists and ranchers have fought over the heart of the American West—the wide-open spaces that stretch from our cities to the "purple mountain majesties" we sang of in school.

The combatants have fought long and hard, but as their struggle over the working landscapes of the West pulled in citizens, agency officials, attorneys, and judges, one consequence is clear: during the fight, millions of acres of the West's open spaces and biologically rich lands were broken by development.

There have been other unintended consequences. Forest Service and Bureau of Land Management officials who once physically managed our purple mountain majesties now mostly manage mountains of paper. Endangered species hang on by claw or beak despite hundreds of lawsuits. Rural towns simply hang on.

Meanwhile, human communities divide into factions. Most tragically, the stewards of working landscapes are surrendering their lands at unprecedented rates to the pressure that tears the quilt of nature into rags.

Perhaps the fight had to happen. The West's grasslands and streams and wildlife were in trouble from a century or more of hard use when this fight was joined. The nation had to debate the use of 420,000 square miles of grazed public land across eleven states. But the fight has gone on far too long. In recent years, the American West has witnessed tremendous positive changes, including the rise of models of sustainable use of public and private lands, the shift of conservation and scientific strategies from protection alone to include restoration, and the expanding role of cooperative efforts to move beyond resource conflicts.

As a consequence of these crises and trends, we believe it is time to cease hostilities and enter a new era of cooperation.

We believe that how we inhabit and use the West today will determine the West we pass on to our children tomorrow, that preserving the biological diversity of working landscapes requires active stewardship, and that under current conditions, the stewards of those lands are compensated for only a fraction of the values their stewardship provides.

We know that poor management has damaged land in the past and in some areas continues to do so, but we also believe appropriate ranching practices can restore land to health. We believe that some lands should not be grazed by livestock but also that much of the West can be grazed in an ecologically sound manner. We know that management practices have changed in recent years, ecological sciences have generated new and valuable tools for assessing and improving land, and new models of sustainable use of land have proved their worth.

Finally, we believe that the people of the West must halt the further conversion of working landscapes to uses that destroy this wellspring of ecological, aesthetic, and cultural richness that is celebrated around the world.

Time is short. The cost of delay is further irrevocable loss.

We therefore reject the acrimony of past decades that has dominated debate over livestock grazing on public lands, for it has yielded little but hard feelings among people who are united by their common love of land and who should be natural allies.

And we pledge our efforts to form the "Radical Center," where:

- The ranching community accepts and aspires to a progressively higher standard of environmental performance;
- the environmental community resolves to work constructively with the people who occupy and use the lands it would protect;

- the personnel of federal and state land-management agencies focus not on the defense of procedure but on the production of tangible results;
- the research community strives to make their work more relevant to broader constituencies;
- the land-grant colleges return to their original charters, conducting and disseminating information in ways that benefit local landscapes and the communities that depend on them;
- the consumer buys food that strengthens the bond between his or her own health and the health of the land;
- the public recognizes and rewards those who maintain and improve the health of all land;
- and that all participants learn better how to share both authority and responsibility.

As the ranks of the Radical Center swell with those who are committed to these goals, the promise increases that "America the Beautiful" may become an image of the future as well as of the past and, with the grace of good fortune, the West may finally create what Wallace Stegner called "a society to match its scenery."

In the expectation that we face a better future for the West, we hereby sign our names and invite others to add their own:

- Michael Bean, conservationist, Environmental Defense
- Jim Brown, ecologist, University of New Mexico
- Bob Budd, manager of Red Canyon Ranch for the Nature Conservancy
- Bill deBuys, author, conservationist
- Kris Havstad, supervisory scientist at the USDA ARS/Jornada Experimental Range
- Paul Johnson, former chief of the Natural Resources Conservation Service
- Teresa Jordan, author
- Daniel Kemmis, Center for the Rocky Mountain West
- Rick Knight, professor of wildlife biology, Colorado State University
- Heather Knight, the Nature Conservancy
- Merle Lefkoff, mediator
- Bill McDonald, rancher and executive director of the Malpai Borderlands Group

- Guy McPherson, ecologist, University of Arizona
- Ed Marston, journalist and former publisher of High Country News
- Gary Paul Nabhan, author and director of the Center for Sustainable Environments, Northern Arizona University
- Duke Phillips, rancher,
- Nathan Sayre, anthropologist
- Paul Starrs, professor of geography, University of Nevada, Reno
- Bill Weeks, the Nature Conservancy
- Courtney White, the Quivira Coalition

SEVENTEEN

HOPE ON THE RANGE

(2004)

NEAR BROTHERS, CENTRAL OREGON

DOC AND CONNIE Hatfield like circles. When they give a talk, they often ask the audience to sit in a big circle, so everyone can see one another. Their goal is to encourage participation, which is why they literally refuse to be the center of attention. Circles, they believe, create a feeling of being a part of a large family.

Which is a fair description of Oregon Country Beef (OCB), the food cooperative that Doc and Connie founded in 1976 (known today as Country Natural Beef).

Yet the conversation in a Hatfield circle is hard-nosed and economic-minded as well, which also describes OCB. The frank talk focuses on profit, healthy food, markets, marketing, progressive management, and bankers. They speak from experience, and they have a success story to tell. Bankers love Oregon Country Beef, they tell the circle. So do its customers. So when Connie tells the ranchers in the room to "decommodify or die," as she invariably does, the circle listens closely.

In the mid-1980s, the Hatfield family ranch was broke and going out of business. Nothing was working right—beef prices were low, pressure from environmentalists was high, profits were nonexistent, and hope was fading. Desperation ruled, and not just on the Hatfields' place. All across central and eastern Oregon, neighbors and friends on ranch after ranch were struggling to hang on economically and emotionally. Clearly, business as usual was failing.

Fast-forward eighteen years. Today, the situation has been completely reversed. In place of despair, hope rules the range.

"My favorite indicator," says Connie, "is how many babysitters we need at our annual meeting to watch the little ones. In the beginning, we didn't need a single one. Today, we need three."

That's because Oregon Country Beef has grown from fourteen participating ranches to seventy. Families are not only staying put and making a living; some have returned home from distant points. There are other indicators. A discriminating consumer can find Oregon Country Beef in grocery stores from Fresno, California, to Bellingham, Washington, to Boise, Idaho. The market for its locally grown, natural beef continues to expand. In fact, OCB struggles at times to keep up with demand.

"We could add another twenty ranches easy," says Connie. "But we're kinda picky. Not everyone who wants in can adjust to our model. We make decisions by consensus, for instance. That means giving up some cherished independence, which is hard for ranchers. But that's what we do."

Not everyone, in other words, likes to sit in circles.

Oregon Country Beef was born in 1986, when Connie Hatfield, driven to desperation, decided to confront her nemesis. She drove from her ranch near Brothers, in central Oregon, forty-five miles west to Bend, the biggest city in the area. She wasn't going to confront an anti-grazing environmentalist, however, or a federal bureaucrat. Instead, she confronted a health guru.

"I went into a fitness center and asked the owner what he thought about red meat," she recalls. "To my surprise, he told me he loved red meat. In fact, he ate it three times a week. But he wanted healthy meat, which meant he had to buy it from Argentina! That's because it didn't have any hormones or antibiotics in it."

Connie quickly saw two marketing opportunities. "First, we could produce a healthy product for the consumer, and second, it could be local," she says. "They fit together perfectly."

When Connie began to ask around, she found fourteen ranch families willing to give the idea of OCB a try. Together, they made some early critical decisions about membership:

- The meat would be certified "natural"—free of antibiotics, steroids, hormones, and other chemicals.
- Each family would give at least ten days a year to group meetings, as well as at least one day greeting customers at stores in Portland and other cities.
- Ranches would be available to host tours for meat buyers.
- Each ranch would abide by third-party certification standards for land stewardship.
- Each ranch would help OCB provide a year-round (fresh) product.
- Each ranch would craft a set of goals to describe the sorts of lives they wished to lead, the desired condition of their land and livestock, the type of product they strove to produce, and the actions they would take to achieve those goals.

It was a conscious departure from the "branded beef" programs pushed by the major food corporations, which often simply promote one type of animal, such as Angus, over another. The Hatfields weren't buying this strategy.

"Consumers today want to know what's in their food, where it came from, and what's happening to the land," says Connie. "But they're busy too, and they often don't have time think about the details. They want to do the right thing, but they often don't know what that means."

After nineteen years of feedback, the Hatfields have discovered that taste is the consumer's number one concern. "They want a product that is fresh and tastes good," says Connie. "That's why they come back."

The issue of sustainable stewardship, however, remains strong for OCB ranchers. Over the years, they have developed a set of management principles they call "Grazing Well," to which all participating ranches conform. They include:

- proper water cycling: dense stands of perennial plants, grass litter on the ground, and native shrubs in the riparian areas—all capturing and holding water.
- using rotational grazing of livestock so that grasses are given time to recover, including the deferment of pastures year to year.
- employing low-stress livestock-handling methods.

- maintaining biodiversity, including predators, birds, and other wildlife.
- planning for long-term health rather than short-term maximization of resources.

Still, for all the goals and principles that make OCB unique, the bottom line is top priority. Doc Hatfield put it this way: "You've got to make money every month or you're not doing something right."

And it's all done with a handshake.

"We had $25 million in boxed beef sales without a written contract," says Connie. "It's all based on trust and honor." Best of all, seven young families have returned to their ranches. That's because they can make a living in the beef business now. Things have gone so well, in fact, OCB isn't taking on new customers, preferring instead to concentrate on expanding their base. "Unlike other meat operations," says Connie, "we decided we needed a lot of space in a few stores, not a little space in a lot of stores. And that's worked well for us."

According to Doc, a major key to profitability is forecasting. OCB plans eighteen months in advance, guaranteeing a price to the ranchers, free from commodity market fluctuations. Each producer has a good projection of what they will get, and when, for their cattle.

OCB members also control the animals from birth to slaughter. A typical OCB animal spends the first eighteen months of its life on grass. Then it is moved to a family-run feedlot for ninety days before being shipped out. There is no animal fat or blood in the feed, and if an animal needs anything beyond routine vaccinations, it is removed from the program.

Another key is to know the real cost of production, including long-term ecological sustainability.

"Most ranchers have no idea of their true costs," says Doc. "They know what their bills are, but they have no idea about the value of their land over time. The traditional cost of production on a ranch is only what it takes to produce a pound of beef. We include the larger ecological costs, blended into a package and marketed as a whole."

When a ranch is ecologically healthy and economically sustainable, Doc says, "you have a perpetual-motion economic engine."

NEAR SENECA,
EASTERN OREGON

"A SIGN OF a true Western town is its honorable poverty," says Jack Southworth, describing his tiny hometown of Seneca, in east-central Oregon.

The Southworth brothers' ranch was one of the fourteen original OCB ranches, and Jack remains an active participant. He drafted the ecological stewardship guidelines by which the Food Alliance, a nonprofit organization based in Portland, certifies each operation. He also continues to volunteer as a facilitator at the regular meetings of OCB members.

Southworth credits his involvement in OCB with turning his ranch, and his life, around. Economically, the fixed price he gets for his cattle gives him a critical degree of financial security and allows him to plan ahead more effectively. "I stopped trying to hit home runs every time and focused on hitting singles instead," he says. "That's helped a lot. We don't get the highest prices this way, but we avoid the downturns too."

Ecologically, OCB's emphasis on good stewardship dovetails with the close attention Jack pays to the health of his land. Socially, OCB membership has created a sense of family that has gone a long way to reduce stress in Jack's life. Overall, OCB enables ranchers like the Southworths to give something back—to the community, to the region, and to the land. "It's not just the food; it's the connection with the customers that I enjoy," says Jack. "They give you a sense of well-being that I never got from the commodity market."

For all of his financial security, Jack Southworth may be most proud of his willows. Healthy, dense stands line both sides of Silvies River, which meanders across the ranch. It didn't look like this when Jack was growing up. In fact, he remembers using a tractor to pull the very last willow clump out of the ground, under orders from his father, when he was twelve.

"My father wanted grass right to the edge of the water and nothing else," Jack recalls. "The trouble was, that's not what the river wanted. Soon we had a big problem."

Without adequate vegetative protection, the riverbanks began to erode. Alarmed, his father began to deposit old cars in the water in a desperate attempt to stem the erosion. It didn't work. When Jack took over the ranch right out of college, he tried a different strategy. He decided to plant willows and fence the cows out.

His father wasn't at all pleased. "My dad was a tough old World War II Marine, and he was pretty well set in his ways," says Jack. "Maybe it was a generational thing. Dad tried to control the land. My approach is to go with what nature gives you."

Toward that end, Jack and his wife wrote out a three-part goal statement for their ranch. The first two parts focus on community and livestock well-being. The third reads: "To bring about the quality of life and products we desire, we need a dense stand of perennial grasses with some shrubs. We want the ground between plants to be covered with decaying plant litter. We want the streams to be lined with willows, home to beaver, and good habitat for trout. We want the precipitation we receive to stay on the ranch as long as possible and to leave here as late-season stream flows or plant growth."

And they've done just that. A recent inspection report by the Food Alliance noted the following accomplishments on the Southworth ranch:

- "Livestock are grazed to maintain and enhance perennial plant communities and spread manure over the ground."
- "The manager does not use herbicide weed control. The manager uses cattle to reduce plant vigor and seed production of problem plants, while promoting the growth of desirable vegetation to compete with weeds. No weed problems were observed on this ranch."
- "Ranch management has resulted in improved riparian areas and upland vegetation. Willows have been planted along streams to improve the diversity of riparian communities and improve bank stability, benefiting both fish and wildlife."
- "Great care was [taken to explain] how stress is kept low for animals and people."
- "Manager is continually trying new things, evaluating the results and making improvements on the ranch . . . The Food Alliance has no substantive comments to offer, your scores are exemplary."

NEAR STEENS MOUNTAIN,
SOUTHEASTERN OREGON

AT THE OTHER end of the scale, at least superficially, is the Roaring Springs Ranch, another OCB member. The Southworth ranch is relatively small in size; the Roaring Springs Ranch, located on the flanks of Steens Mountain in southeast Oregon, is large. Its acreage, combined with a nearby ranch, runs over six hundred thousand acres, making it one of the largest operations in the state. The owners and manager sometimes use a helicopter to get around.

On closer inspection, however, the similarities between the two OCB ranches are more striking than their differences. That's because both operations aim for the same goal: progressive management in service of human, animal, and ecological health. And both achieve this goal through visionary and energetic leadership.

On the Roaring Springs, the leadership is provided by Stacy Davies, a studious former employee of Doc and Connie Hatfield. Davies runs over four thousand head of mother cows on the ranch, providing a large part of OCB's annual supply of animals. In doing so, he earns a comfortable living for himself and his family, including his wife Elaine and six sons—thus fulfilling Connie Hatfield's principal criteria for success.

Stacy calculates that he can wean a calf at a cost of sixty cents per pound, thanks to low inputs, low labor costs, and good grass. He insists that the employees earn decent salaries and enjoy a good quality of life, so mostly he focuses on lowering other costs of production. "If I could, I'd park every machine on the ranch and never start another engine," he says.

Like Jack Southworth, Davies believes that making the management fit the land increases profit.

"For the Roaring Springs, the best use of our natural resources is an April calf," he says. "This way, there aren't any conflicts with predators, labor costs are lower, and I can still wean a 450-pound calf in the fall."

Instead of supplementing his calves through the winter, Davies ships them to California for green grass and then brings them back in May for more grass. This way, he can supply eight-hundred-pound feeders to OCB eight months of the year. He doesn't object to feedlots because he believes the consumer demands consistency in the meat—something that's much harder to control with grass-only animals. At the same time, he's no fan of government incentives. He thinks the market should determine who gets paid and how much, which is why he likes the OCB model.

"If it's truly important to the American people, then they should pay for it directly," he says.

Davies plows a significant portion of the ranch's profits into conservation. He does so for a number of reasons, not the least of which is maintenance of profitability. He calls it a "reinvestment" in the ranch's long-term health. For example, he pays a crew fifty dollars an acre to clear the abundant juniper trees, which he considers to be "big weeds," on the ranch's private land. His concern over juniper is a familiar story across the West: the suppression of natural fire over the decades has resulted in an explosion of woody vegetation and a diminishment of historic grasslands. The difference on the Roaring Springs is that Davies has the means, and the desire, to act.

However, in a move that typifies the Roaring Springs, Davies acts in a manner that is at once innovative and frugal. Rather than cut and stack the junipers for eventual burning, Davies has his crew skid the trees into large, circular windrows that act as cattle exclosures for pastures that need rest or recovery. These "fences" cost him just $1,200 a mile to construct, compared to $4,500 per mile for barbed wire. When the exclosures are no longer needed, he lights a fire and burns them up.

The reintroduction of fire, in fact, is a big part of Davies's conservation mission on the Roaring Springs. So is wildlife, which abounds. Populations of antelope and bighorn sheep dot the ranch, as do herds of wild horses. Sage grouse, a species in peril across the region, flourish on the Roaring Springs, Davies says. He believes his sage grouse populations are healthier than those on two nearby national wildlife refuges. Next year, he plans to hire a full-time wildlife biologist to help him understand better the dynamics at work.

Clearly, Stacy Davies, like Jack Southworth and the Hatfields, enjoys a challenge—even thrives on it—including the challenge of setting high standards and then meeting them. He also likes to set precedents.

A major opportunity to do the latter came his way in the late 1990s when Bruce Babbitt, then secretary of the interior, publicly considered creating a national monument on Steens Mountain. A classic recipe for conflict was set in motion: urban environmentalists wanted the monument designation to protect the mountain, while the local residents wanted to be left alone. Protect it from what, they wondered?

After a lengthy, and sometimes testy, process of dialogue and wheeling and dealing, a compromise was brokered. No official monument designation was made. Instead, the upper part of Steens was designated as the first official "cattle-free wilderness" in the nation. At the same time, local ranchers, including the owners of the Roaring Springs, were able to consolidate their private holdings by swapping land with the government. Neither side was completely happy, but it could have been worse. Stacy Davies was in the thick of the negotiations from the start. Characteristically, he understates the conflict as a learning experience—whose principle lesson has a message for us all.

"What I learned was this," he said. "Society needs a goal statement."

EIGHTEEN

THE WORKING WILDERNESS

(2005)

"The only progress that counts is that on the actual landscape of the back forty."

—ALDO LEOPOLD

U BAR RANCH

SILVER CITY, NEW MEXICO

DURING A CONSERVATION tour of the well-managed U Bar Ranch near Silver City, New Mexico, I was asked to say a few words about a map a friend had recently given to me.

We were taking a break in the shade of a large piñon tree, and I rose a bit reluctantly (the day being hot and the shade being deep) to explain that the map was commissioned by an alliance of ranchers concerned about the creep of urban sprawl into the five-hundred-thousand-acre Altar Valley, located southwest of Tucson, Arizona. What was different about this map, I told them, was what it measured: indicators of rangeland health, such as grass cover (positive) and bare soil (negative), and what they might tell us about livestock management in arid environments.

What was important about the map, I continued, was what it said about a large watershed. Drawn up in multiple colors, the map expressed the intersection of three variables: soil stability, biotic integrity, and hydrological function—soil, grass, and water, in other words. The map displayed three conditions for each variable—"Stable," "At Risk," and "Unstable"—with a color representing a particular intersection of conditions. Deep red designated an unstable, or unhealthy, condition for soil, grass (vegetation), and water, for example, while deep green represented stability in all three. Other colors represented conditions between these extremes.

In the middle of the map was a privately owned ranch called the Palo Alto. Visiting it recently, I told them, I had been shocked by its condition. It had been overgrazed by cattle to the point of being nearly "cowburnt," to use author Ed Abbey's famous phrase. As one might expect, the Palo Alto's color on the map was blood red, and there was plenty of it.

I paused briefly—now came the controversial part. This big splotch of blood red continued well below the southern boundary of the Palo Alto, I said. However, this was not a ranch, but part of the Buenos Aires National Wildlife Refuge, a large chunk of protected land that had been cattle-free for nearly sixteen years.

That was as far as I got. Taking offense at the suggestion that the refuge might be ecologically unfit, a young woman from Tucson cut me off. She knew the refuge, she explained, having worked hard as a volunteer with an environmental organization to help "heal" it from decades of abuse by cows.

The map did not blame anyone for current conditions, I responded; nor did it offer opinions on any particular remedy. All it did was ask a simple question: Is the land functioning properly at the fundamental level of soil, grass, and water? For a portion of the Buenos Aires National Wildlife Refuge, the answer was "no." For portions of the adjacent privately owned ranches, which were deep green on the map, the answer was "yes."

Why was that a problem?

I knew why. I strayed too closely to a core belief of my fellow conservationists—that protected areas, such as national parks, wilderness areas, and wildlife refuges, must always be rated, by definition, as being in better ecological condition than adjacent "working" landscapes.

Yet the Altar Valley map challenged this paradigm at a basic level, and when the tour commenced again on a ranch that would undoubtedly encompass more deep greens than deep reds on a similar map, I saw in the reaction of the young activist a reason to rethink the conservation movement in the American West.

From the ground up.

CS RANCH
CIMARRON, NEW MEXICO

MY DECISION RECEIVED a boost a few weeks later while sitting around a campfire after a tour of the beautiful one-hundred-thousand-acre CS Ranch located in northeastern New Mexico. Staring into the flames, I found myself thinking about ethics. I believed at the time, as do many conservationists, that the chore of ending overgrazing by cattle in the West was a matter of getting ranchers to adopt an ecological ethic along the lines that Aldo Leopold suggested in his famous essay "The Land Ethic," where he argued that humans had a moral obligation to be good stewards of nature.

The question, it seemed to me, was how to accomplish this lofty goal.

I decided to ask Julia Davis-Stafford, our host, for advice. Years earlier, Julia and her sister Kim talked their family into switching to holistic management of the land, a decision that over time caused the ranch to flourish economically and ecologically. In fact, the idea for my query came earlier that day when I couldn't decide which was more impressive: the sight of a new beaver dam on the ranch or Julia's strong support for its presence.

The Davis family, it seemed to me, had embraced Leopold's land ethic big time. So, over the crackle of the campfire, I asked Julia, "How do we get other ranchers to change their ethics too?"

Her answer altered everything I had been thinking up until that moment.

"We didn't change our ethics," she replied. "We're the same people we were fifteen years ago. What changed was our knowledge. We went back to school, in a sense, and we came back to the ranch with new ideas."

Knowledge *and* ethics, neither without the other, I suddenly saw, are the key to good land stewardship. Her point confirmed what I had observed during visits to livestock operations across the region: many ranchers *do* have an environmental ethic, as they have claimed for so long. Often their ethic is a powerful one. But it has to be matched with *new* knowledge—especially ecological knowledge—so that an operation can adjust to meet changing conditions, both on the ground and in the arena of public opinion. Of course, a willingness on the part of a rancher to "go back to school" is a prerequisite to gaining new insights. Tradition, however, seemed to have a lock on many ranchers.

The same thing is true of many conservationists. In the years since I cofounded the Quivira Coalition, I came to the conclusion that it had been a long time since any of us had been back to school ourselves. Tradition was just as much an obstacle in the environmental community as it was in agriculture. It wasn't just the persistence of various degrees of bovine bigotry among activists, despite examples of healthy, grazed landscape like the U Bar, either. It was more a stubbornness about the relation between humans and nature—they should be kept as far apart as possible—expressed in the long-standing dualism of environmentalism that said recreation and play in nature were preferable to work and use.

If conservationists went back to school, as the Davis family did, what could we learn? Aldo Leopold had a suggestion that can help us today: study the fundamental principle of *land health*, which he described as "the capacity of the land for self-renewal," with conservation being "our effort to understand and preserve this capacity."

By studying the elements of land health, especially as they change over time, conservationists could learn that grazing is a natural process. The consumption of grass by ungulates in North America has been going on for millions of years—not by cattle, of course, but by bison, elk, and deer (and grasshoppers, rabbits, and even ants)—resulting in a complex relationship between grass and grazer that is ecologically self-renewing. We could learn that a re-creation of this relationship with domesticated cattle lies at the heart of the new ranching movement, which is why many progressive ranchers think of themselves as "grass farmers" instead of beef producers.

We could also learn that many landscapes need periodic pulses of energy, in the form of natural disturbance—such as fires and floods (but not the catastrophic kind)—to keep things ecologically vibrant. Many conservationists know that low-intensity fires are a beneficial form of disturbance in ecosystems because they reduce tree density, burn up old grass, and aid nutrient cycling in the soil. But many of us don't know that small flood events can be positive agents of change too, as can drought, windstorms, and even insect infestation. Or that animal impact caused by grazers, including cattle, can be a beneficial form of disturbance.

We could further learn, as the Davis family did, that the key to healthy disturbance with cattle is to control the timing, intensity, and frequency of their impact on the land. The CS, and other progressive ranches, bunch their cattle together and keep them on the move, rotating the animals frequently through numerous pastures. Ideally, under this system, no single piece of ground is grazed by cattle more than once a year, thus ensuring plenty of time for the plants to recover. The keys are regulating where cattle go, which can be done with fencing or a herder, and the timing of their movement, in which the herd moves are carefully planned and monitored. In fact, as many ranchers have learned, overgrazing is more a function of timing than it is of numbers of cattle. For example, imagine the impact 365 cows would have in one day of grazing in one small pasture versus what one cow would do in 365 days of grazing in the same pasture. Which is more likely to be overgrazed? Hint: have you ever seen what a backyard lot looks like after a single horse has grazed it for a whole year?

We could also learn, as I did, that much of the damage we see today on the land is historical—a legacy of the "boom years" of cattle grazing in the West. Between 1880 and 1920, millions of hungry animals roamed uncontrolled across the range, and the overgrazing they caused was so extensive, and so alarming, that by 1910, the U.S. government was already setting up programs to slow and to heal the damage. Today, cattle numbers are down, way down, from historic highs—a fact not commonly voiced in the heat of the cattle debate.

A willingness to adopt new knowledge allowed the Davis family to maintain their ethic yet stay in business. Not only did it improve their bottom line; it also helped them meet evolving values in society, such as a rising concern among the pubic about overgrazing. Rather than fight change, they had switched.

As the embers of the campfire burned softly into the night, I wondered if the conservation movement could do the same.

KAIBAB NATIONAL FOREST
FLAGSTAFF, ARIZONA

A FRIEND OF mine likes to tell a story about the professor of environmental studies he knows who took a group of students for a walk in the woods near Flagstaff, Arizona. Stopping in a meadow, the professor pointed at the ground and asked, not so rhetorically, "Can anyone tell me if this land is healthy or not?" After a few moments of awkward silence, one student finally spoke up and said, "Tell us first if it's grazed by cows or not." In a similar vein, a Santa Fe lawyer told me that a monitoring workshop at the boundary between a working ranch and a wildlife refuge south of Albuquerque had completely rearranged his thinking. "I've done a lot of hiking and thought I knew what land health was," he said, "but when we did those transects on the ground on both sides of the fence, I saw that my ideas were all wrong."

These two instances illustrate a recurring theme in my experience as a conservationist. To paraphrase a famous quote by a Supreme Court justice, members of environmental organizations can't define what healthy land is, but they know it when they see it.

The principle problem is that we are "land illiterate." When it comes to "reading" a landscape, we might as well be studying a foreign language. Many of us who spend time on the land don't know our perennials from our annuals, what the signs of poor water cycling are, what leads to a deeply eroded gully, or, simply by looking, whether a meadow is healthy or not.

For a long time, this situation wasn't our fault. What all of us lacked—rancher, conservationist, range professional, curious onlooker—was a common language to describe the common ground below our feet. But that has changed.

In recent years, range ecologists have reached a consensus on a definition of health: the degree to which the integrity of the soil and ecological processes of rangeland ecosystems are sustained over time. These include water and nutrient cycling, energy flow, and the structure and dynamics of plant and animal communities. In other words, when scarce resources such as water and nutrients are captured and stored locally, by healthy grass plants, for example, then ecological integrity can be maintained and sustained. Without them—if water runs off-site instead of percolating into the soil, or grass plants die due to excessive erosion of the topsoil, for example—this integrity will likely be lost over time, perhaps quickly.

This is the language of soil, grass, and water.

Taking it to the next step, range ecologists echo Aldo Leopold's famous quote that "Healthy land is the only permanently profitable land." Producing commodities and satisfying values from a stretch of land on a sustained basis, they insist, depends on the renewability of internal ecological processes. In other words, before land can sustainably support a value, such as livestock grazing, hunting, recreation, or wildlife protection, it must be functioning well at a basic ecological level. Before we, as a society, can talk about designating critical habitat for endangered species, or increasing forage for cows, or expanding recreational use, we need to know the answer to a simple question: is the land healthy at the level of soil, grass, and water?

If the answer is "no," then all our values for that land may be at risk.

Or as Kirk Gadzia, an educator, range expert, and coauthor of *Rangeland Health,* the pioneering 1994 book published by the National Academy of Sciences, likes to put it, "It all comes down to soil. If it's stable, there's hope for the future. But if it's moving, then all bets are off for the ecosystem." It is a sentiment Roger Bowe, an award-winning rancher from eastern New Mexico, echoes. "Bare soil is the rancher's number one enemy."

It should become the number one enemy of conservationists as well.

The publication of *Rangeland Health* was the touchstone for a new consensus on the meaning of land health within the scientific and range professional communities. It paved the way for the debut, in 2000, of a federal publication entitled *Interpreting Indicators of Rangeland Health*, which provides a seventeen-point checklist for the qualitative assessment of upland health. A similar assessment has been made of stream health by a federal interagency group known as the National Riparian Team. The indicators of health include measures of the presence of rills, gullies, bare ground, pedestaling (grass plants left high and dry by water erosion), litter (dead grass, which retards the erosive impact of rain and water), soil compaction (which can prohibit water infiltration), plant diversity (generally a good thing), and invasive species (generally not)—the same indicators that formed the basis of the Altar Valley map that I described on the tour.

This was the message I tried to communicate to the young activist under the tree that hot summer day—that a rangeland health paradigm, employing standard indicators, allows all land to be evaluated equally and fairly. By adopting it, the conservation movement could begin to heed Aldo Leopold's advice that any activity that degrades an area's "land mechanism," as he called it, should be curtailed or changed, while any activity that maintains, restores, or expands it should be supported. It should not matter if that activity is ranching or recreation.

CHACO NATIONAL HISTORICAL PARK
SOUTHEAST OF FARMINGTON, NEW MEXICO

IN AN ATTEMPT to understand the issues of land health better, I paid a visit to a famous fence-line contrast. This particular fence separated the Navajo Nation, and its cows, from Chaco Culture National Historical Park, a UNESCO World Heritage site and archaeological preserve located in the high desert of northwest New Mexico. Cattle-free for over fifty years, Chaco's ecological condition became a pedagogical issue some years ago when Allan Savory used the boundary to highlight the dangers in the park of too much rest from the effects of natural disturbance, including grazing and fire.

I wasn't a fan of fence-line contrasts myself, mostly because I dislike dichotomies represented by a fence: us/them, either/or, wild/unwild, grazed/ungrazed. The world is more complicated than that. I'd rather take fences down, or move beyond them. But fence-line contrasts have pedagogical value, especially for new students of range health—like me. I decided I wanted to see this contrast in particular, but I knew I needed help interpreting what I saw, so I asked Kirk Gadzia to come along.

Both of us were well aware of the park's history—that a century of overgrazing by livestock had badly degraded the land surrounding the famous ruins. We also understood that the era's typical response to this legacy of overuse was to protect the land from further degradation with the tools of federal ownership and a barbed-wire fence. That's how Chaco became a national park. At the time, it was a common and appropriate scenario played out all across the West. But Kirk and I didn't go to Chaco to argue with history or to pick a fight with the National Park Service. We weren't there to offer solutions to any particular problem either. We simply wanted to take the pulse of the land on both sides of a fence.

We stopped along the road at the eastern boundary of the park (this was during the growing season). On the Chaco side, we saw a great deal of bare ground, as well as many forbs, shrubs, and other woody material, some of it dead. We saw few young plants, few perennial or bunch grasses, lots of wide spaces between plants, lots of oxidized plant matter (dead grass turning gray in the sunlight), and a great deal of poor plant vigor. We saw both undisturbed, capped soil (bad for seed germination) and abundant evidence of soil movement, including gullies and other signs of erosion. On the positive, we saw a greater diversity of plant species than on the Navajo side, more birds, more seed production, no sign of manure, and no sign of overgrazing.

On the Navajo side, we saw lots of plant cover and litter, lots of perennial grasses, tight spaces between plants, few woody species, a wide age-class distribution among the plants, little evidence of oxidization, and lots of bunch grasses. We saw little evidence of soil movement, no gullies, and far fewer signs of erosion than on the Chaco side. On the other hand, we saw less species diversity, poor plant vigor, a great deal of compacted soil, fewer birds, less seed production, a great deal of manure, and numerous signs of overgrazing.

"So, which side is healthier?" I asked Kirk.

"Neither one is healthy, really," he replied, "not from a watershed perspective anyway." He noted that the impact of livestock grazing on the Navajo side was heavy; plants were not being given enough time to recover before being bitten again (Kirk's definition of overgrazing). As a result, the plants lacked the vigor they would have exhibited in the presence of well-managed grazing.

However, Kirk thought the Chaco side was in greater danger, primarily because it exhibited major soil instability due to gullying, capped soil, and lack of plant litter. "The major contributing factor to this condition is the lack of tightly spaced perennial plants," he continued, "which exposes the soil to the erosive effects of wind and rain. When soil loss is increased, options for the future are reduced."

"But isn't Chaco supposed to be healthier because it's protected from grazing?"

"That's what people always seem to assume," said Kirk. "In my experience in arid environments around the world, total rest from grazing has predictable results. In the first few years, there is an intense response in the system as the pressure of overgrazing is lifted. Plant vigor, diversity, and abundance often return at once, and all appears to be functioning normally. Over the years, however, if the system does not receive periodic natural disturbance, by fire or grazing, for example, then the overall health of the land deteriorates. And that's what we are seeing on the Chaco side."

Then he added a caveat.

"Maybe land health isn't the issue here," he said. "It may be more about values. Is rest producing what the park wants? Ecologically, the answer is probably 'no.' But from a cultural perspective, the answer might be 'yes.' From the public perspective too. People may not want to see fire or grazing in their park."

But at what price, I wondered? Later in the day, we learned that the Park Service was so worried about the threat of erosion to Chaco's world-class ruins that they intended to spend a million dollars constructing an erosion-control structure in the Chaco Wash. This told us the agency knows it has a "functionality" crisis on its hands.

But how can proper functioning condition be restored if the Park Service's hands are tied by a cultural value that says Chaco must be protected from incompatible activities, even those that might have a beneficial role to play in restoring the park to health?

As I drove home, I realized that this tension between "value" and "function" at Chaco was sign of a new conflict spreading slowly across the West—symbolized by a fence. The cherished "protection" paradigm, embedded in the conservation movement since the days of John Muir, rubbed against something new, something energetic—something beyond the fence.

BANDELIER NATIONAL MONUMENT

NEAR LOS ALAMOS, NEW MEXICO

THE PASSAGE OF the Wilderness Act in 1964 was a seminal event in the history of the American conservation movement. For the first time, wilderness had a legal status, enabling the designation and the protection of "wildland," which had been under siege in that era of environmental exploitation. Energized, the conservation movement grabbed the wilderness bull by both horns and has not let go to this day. But the act's passage also had an unforeseen consequence—it set in motion the modern struggle between value and function in our Western landscapes.

This tension took a while to develop. In 1964, there was intellectual harmony between the social and ecological arguments for the creation of a federal wilderness system. No reconciliation was necessary between the act's definition of wilderness as a tract of land "untrammeled by man . . . in which man is a visitor who does not remain" and Aldo Leopold's declaration, published in *A Sand County Almanac* fifteen years earlier, that wilderness areas needed protection because they were ecological "base datums of normality."

Leopold asserted that wilderness was "important as a laboratory for the study of land health," insisting that in many cases, "we literally do not know how good a performance to expect of healthy land unless we have a wild area for comparison with sick ones." Author Wallace Stegner extended the medical metaphor when he argued that wilderness was "good for our spiritual health even if we never once in ten years set foot in it."

But a lot has changed in the years since the passage of the Wilderness Act. While most Americans still believe wilderness is necessary for social and mental health, few ecologists now argue that wilderness areas can be considered as "base datums" of ecological health.

For example, in an article published in the journal *Wild Earth* in 2001, entitled "Would Ecological Landscape Restoration Make the Bandelier Wilderness More or Less of a Wilderness?" the authors, including ecologist Craig Allen, who has studied Bandelier National Monument, located in north-central New Mexico, for nearly twenty years, state matter-of-factly that "Most wilderness areas in the continental United States are not pristine, and ecosystem research has shown that conditions in many are deteriorating."

In their opinion, the Bandelier Wilderness is suffering from "unnatural change" as a result of historic overuse of the area in the late nineteenth and early twentieth centuries—grazing by sheep principally—which triggered unprecedented change in the park's ecosystems, resulting in degraded and unsustainable conditions. "Similar changes," they write, "have occurred throughout much of the Southwest."

Specifically, soils in Bandelier are "eroding at net rates of about one-half inch per decade. Given soil depths averaging only one to two feet in many areas, there will be loss of entire soil bodies across extensive areas." This is bad because the loss of topsoil, and the resulting loss of water available for plants, impedes the growth of all-important grass cover, thus reducing the incidence of natural and ecologically necessary fires.

The elimination of livestock grazing with the creation of the park in the 1930s was no panacea for Bandelier's functionality crisis, however. Herbivore exclosures established in 1975 show that protection from grazing, by itself, "fails to promote vegetative recovery." Without management intervention, they argue, this human-caused case of accelerated soil erosion will become irreversible. "To a significant degre, the park's biological productivity and cultural resources are literally washing away."

Their summation is provocative: "We have a choice when we know land is 'sick.' We can 'make believe,' to quote Aldo Leopold, that everything will turn out all right if Nature is left to take its course in our unhealthy wildernesses, or we can intervene—adaptively and with humility—to facilitate the healing process."

I believe new knowledge about the condition of the land leaves us no choice: we must intervene. However, this turns a great deal of old conservation thinking on its head.

For instance, Wallace Stegner once wrote, "Wildlife sanctuaries, national seashores and lakeshores, wild and scenic rivers, wilderness areas created under the 1964 Wilderness Act, all represent a strengthening of the decision to hold onto land and manage large sections of the public domain rather than dispose of them *or let them deteriorate*" (emphasis added).

But we have let them deteriorate—as the Buenos Aires, Chaco, and Bandelier examples demonstrate. Whether their deteriorated condition is a result of historical overuse or some more recent activity is not as important as another question: what are we going to do to heal land we know to be sick?

Clearly it's not 1964 anymore. The harmony between value and function in the landscape, including our protected places, has deteriorated along with the topsoil. This functionality crisis raises important questions for all of us. What, for instance, are the long-term prospects for wildlife populations in the West, including keystone predator species, if the ecological integrity of these special places is being compromised at the level of soil, grass, and water? Also, does protection from human activity preclude intervention, and if so, at what cost to ecosystem health? And on a larger scale, how do we protect our parks and wildernesses from the effects of global warming, acid rain, and noxious weed invasion?

Furthermore, the dualism of protected versus unprotected creates a stratification of land quality and land use that bears little relation to land health. As conservationist Charles Little has written, "Leopold insisted on dealing with land whole: the system of soils, waters, animals, and plants that make up a community called 'the land.' But we insist on discriminating. We apply our money and our energy in behalf of protection on a selective basis." He goes on to say, "The idea of a hierarchy in land quality is *the* tenet of the conservation and environmental movement."

Since John Muir's day, the conservation movement has based this hierarchy on the concept of "pristineness"—the degree to which an area of land remains untrammeled by humans. As late as 1964, when not as much was known about ecology or the history of land use, it was still possible to believe in the pristine quality of wilderness as an ecological fact, as Leopold did. Today, however, pristineness must be acknowledged to be a value, something that exists mostly in the eye of the beholder.

Biologist Peter Raven puts it in blunt ecological terms: "There is not a square centimeter anywhere on earth, whether it is in the middle of the Amazon basin or the center of the Greenland ice cap, that does not receive every minute some molecules of a substance made by human beings."

I believe the new criterion should be *land health*. By assessing land by one standard, a land-health paradigm encourages an egalitarian approach to land quality, thereby reducing conflicts caused by clashing cultural values (theoretically, anyway). By employing land health as the common language to describe the common ground below our feet, we can start fruitful conversations about land use rather than resort to the usual dualisms that have dominated the conservation movement for decades. We can also gain new knowledge about the condition of a stretch of land, and that knowledge can help us make informed decisions.

For example, I know a chunk of Bureau of Land Management (BLM) land west of Taos, New Mexico, that will never be a wilderness area, national park, or wildlife refuge. It is modest land, mostly flat, covered with sage, and very dry. In its modesty, however, it is typical of millions of acres of public land across the West. It is typical in another way too—it exists in a degraded ecological condition, the result of historic overgrazing and modern neglect. A recent qualitative land health assessment revealed its poor condition in stark terms (lots of bare soil, many signs of erosion, and a lack of plant diversity), confronting us with the knowledge that more than forty years of total rest from livestock grazing had not healed the land. Some of it, in fact, teetered on an ecological threshold, threatening to transition to a deeper degraded state.

Fortunately, as humble and unhealthy as this land is, it is not unloved. The wildlife like it, of course, but so do the owners of the private land intermingled with the BLM land, some of whom built homes there. The area's two new ranchers also have great affection for this unassuming land and want to see it healed.

These ranchers are using cattle as agents of ecological restoration. Through the effect of carefully controlled herding, they intend to trample the sage and bare soil, much of which is capped solid (without a cover of grass or litter, soil will often "cap," or seal when exposed to pounding rain, thus preventing seed germination), so that native grasses can get reestablished again.

Using cattle as agents of ecological restoration is not as novel as it may sound. In fact, in his 1933 classic book *Game Management*, Aldo Leopold wrote more generally that wildlife "can be restored with the same tools that have hithertofore destroyed it: fire, ax, cow, gun, and plow." The difference, of course, is the management of the tool, as well as the goals of the tool user.

I believe conservationists should share the same goal as these ranchers: transform red to green on maps such as that of the Altar Valley and the land west of Taos. Whether we use cattle or some other method of restoration, the result must be a thousand acts of healing, starting at the level of soil, grass, and water. And healing must extend to communities of people as well, both urban and rural. Restoration jobs could be a boon to local economies, and volunteers from environmental groups could help. Turning red to green could unite us no matter what our values.

By developing a common language to describe the common ground below our feet, by working collaboratively to heal land and restore rural economies, by monitoring our progress scientifically, and by linking "function" to "value" in a constructive manner, a land-health paradigm can steer us toward fulfilling Wallace Stegner's famous dream of creating a "society to match the scenery."

NINETEEN

THINKING LIKE A CREEK

(2006)

DURING MY TRAVELS, I heard a story about a man who had put short fences across a cattle trail in the sandy bottom of a canyon in Navajo country so that the cattle were forced to meander in an S pattern as they walked, encouraging the water to meander too and thus slow erosion.

I thought this idea was wonderfully heretical. That's because the standard solutions proposed for cattle-caused erosion in creeks were (1) kick the cows out (if you were an environmentalist), (2) ignore it and hope the problem fixes itself (if you were a rancher), or (3) spend a bunch of money on diesel-driven machines and other heavy-handedness (if you worked for an agency). Putting fences in the way of cattle and letting them do the work? How cool.

I learned more about it while attending an environmental restoration conference in downtown Phoenix, of all places. Someone told me that Bill Zeedyk would be giving a talk. I pricked up my ears. "You mean that guy who's been trying to keep part of Hubbell Trading Post from washing away by putting sticks and rocks in the nearby creek bed?" I asked. "The guy who refused to use cement, rip rap, or rock-filled wire baskets?" It was, came the reply. Could this be the same person, I wondered?

It had to be. I decided to catch his talk.

Actually, I bumped into him in the hall not much later. Bill is hard to miss—he looks like a Dutch version of Santa Claus, with ruddy cheeks, twinkling blue eyes, a generous salt-and-pepper beard, and a modest roundness that completes his aura of avuncular charm. Only don't tease Bill too much—as I eventually learned. Bill takes what he does quite seriously.

And what he does is help creeks get better. That might sound like an odd job description, but given the standard environmentalist saw that overgrazing has degraded 80 percent of the region's creeks and riparian areas, compromising their high ecological value in the arid Southwest, strategies of restoration had become an important issue economically, environmentally, and politically over the last decade (hence the conference in Phoenix).

After making quick introductions, I asked Bill if the story about the fences and the cattle trail in the canyon bottom was true. It was, he said. Recognizing that water running down a straight trail will cut a deeper and deeper incision in soft soil with each storm event, increasing the probability of serious erosion trouble, Bill talked the local Navajo ranchers into placing fences at intervals along the trail so that the cows would be forced to create a meander pattern in the soil precisely where Bill thought nature would do so in their absence. Water likes to meander—it's nature's way of dissipating energy—and it will gravitate toward doing so again even if it's temporarily trapped in a cattle-caused rut (or human-caused hiking trail), though it might take a long time. His fence idea was a way to speed up the process, he said.

"What happened after the fences were put it in?" I asked. The water table came up as vegetation grew back, he replied, because the water was now traveling more slowly and had a chance as a result to percolate into the ground, rather than run off like before. Steep, eroded banks began to revegetate as the water table rose, and more water appeared in the bottom of the canyon, which encouraged riparian plant growth.

"Nature did all the heavy lifting," he said, before adding a warm, knowing smile. "It worked too, until someone stole the fences."

I followed him to his talk. Bill's comment reminded me that environmental problems are, as he puts it, "people problems." One is inextricably intertwined with the other. Fixing the environmental problem without addressing the people part, to paraphrase Aldo Leopold, is like fixing the pump without fixing the well.

However, at age seventy, Bill would rather leave the people problem to somebody else. "I'm done arguing," he said to me, "I'd rather focus my energy on fixing creeks and roads."

And that's exactly what he has been doing. Since 1995, five years after his retirement as a biologist with the United States Forest Service, Bill has developed an important set of techniques designed to "heal nature with nature," as I heard in Phoenix that day.

In the presentation, he inventoried this toolbox, illustrating how his low-cost, low-tech methods reduce erosion and sedimentation, return riparian areas to a healthier functioning condition, and restore wet meadows and other wetlands, all at a minimal cost compared to other techniques, such as the backhoe/rock-and-wire gabion-structure approach used by many landowners across the nation.

Bill's toolbox includes:

- one-rock dams (small structures that are literally one-rock high)
- picket baffles and deflectors (wedge-shaped structures that steer water flow)
- wicker weirs (rows of sticks that create a "riffle" effect in creeks)
- vanes (a row of logs pointing upstream that deflect water away from eroding banks)
- headcut control structures (that stop the relentless march of erosion up a creek)
- worm ditches (that redirect water away from headcuts in wet meadows)
- "zuni" rock bowls (small structures that trap water so vegetation can grow)

Many of these structures are placed directly in a watercourse. Vanes and baffles, for instance, often constructed of wooden pickets (harvested locally), are used to deflect stream flow. Weirs are used to control streambed grade and pool depth. One-rock dams are used to stabilize bed elevation, modify slope gradient, retain moisture, and nurture vegetation.

The goal of these structures, I learned, is to stop downcutting in creeks and streams, often by inducing an incised stream to return to a "dynamically stable" channel through the power of small flood events. Bill calls it "Induced Meandering." Its goal is to restore channel dimensions, reestablish appropriate meander patterns and pool/riffle ratios, restore stream access to its floodplain, and raise the water table, which enables riparian vegetation to grow.

In other words, when a creek loses its riparian vegetation—grasses, sedges, rushes, willows, and other water-loving plants—to overgrazing by livestock, say, it tends to straighten out and cut downward because the speed of water is now greater, causing the scouring power of sediment to increase. Over time (and sometimes not very much time), this downcutting results in the creek becoming entrenched below its original floodplain, which causes all sorts of ecological havoc, including a drop in the water table (bad for trees and wet meadows). Eventually, the creek will create a new floodplain at this lower level by remeandering itself, but that is a process that often takes decades.

Bill's idea was to goose the process along by forcing the creek to remeander itself via his vanes, baffles, and riffles, carefully calculated and emplaced. And once water begins to slow down, guess what begins to grow? Willows, sedges, and rushes!

"My aim is to armor eroded streambanks the old-fashioned way," said Bill, "with green, growing plants, not with cement and rock gabions."

The employment of one-rock dams typifies Bill's naturalistic approach. The conventional response of landowners over the years to eroded, downcut streams and arroyos has been to build a check dam in the middle of the watercourse. The old idea was to trap sediment behind a dam, which would give vegetation a place to take root as moisture is captured and stored. Trouble is, check dams work against nature's long-term plans.

"All check dams, big or small, are doomed to fail," said Bill. "That's because nature has a lot more time than we do. As water does its work, especially during floods, the dam will be undercut and eventually collapse, sending all that sediment downstream and making things worse than if you did nothing at all."

"The trick is to think like a creek," he continued. "As someone once told me long ago, creeks don't like to be lakes, even tiny ones. Over time, they'll be creeks again."

One-rock dams, by contrast, don't collapse—because they are only one-rock high. Instead, they slow water down, capture sediment, store a bit of moisture, and give vegetation a place to take root. It just takes more time to see the effect.

"As a species, we humans want immediate results. But nature often has the last word," said Bill. "It took 150 years to get the land into this condition; it's going to take at least as long to get it repaired." The key is to learn how to read the landscape—to become literate in the language of ecological health.

"All ecological change is a matter of process. I try to learn the process and let nature do the work," said Bill, "but you've got to understand the process, because if you don't, you can't fix the problem."

Why even worry about healing creeks in the first place?

For starters, there is a good reason why many authors and historians, including Wallace Stegner, have labeled the American West the "plundered province." More than a century of very hard use, including overgrazing by millions of livestock during the boom years between 1880 and 1920, have created a legacy of damaged, degraded, and just plain worn-out landscapes across the region. Add clear-cut logging, thousands of mines, hundreds of thousands of miles of badly designed and poorly maintained roads, extensive oil-and-gas exploration, and a thousand other cuts from the plundering behavior of individuals and corporations over the years, and you have a region that is chronically in need of a good doctor.

This is hard for most Americans to understand because we spend so much time in national parks, wilderness areas, wildlife refuges, and other pretty places that *seem* healthy. That was certainly my impression growing up in the West. Backpacking through one national park after another, even hiking through the desert as an archaeologist, I had no idea what "land sickness" looked like, to use Leopold's phrase, other than the obvious signs of abuse. That changed when I began Quivira and saw the land health map of the Altar Valley in southern Arizona. But it wasn't until I walked up a deeply eroded arroyo one fine sunny day in 2003 that the magnitude of the problem struck me like a bolt of lightning.

It happened at the boundary between the Gila National Forest and Jim and Joy Williams's ranch, located a few miles south of Quemado, in famously cranky Catron County in west-central New Mexico. We were there as a result of a project we were doing with Bill Zeedyk on Loco Creek, located on the Williams's ranch, which is an ephemeral tributary of Largo Creek, a substantial watercourse in the area.

We had met Jim and Joy Williams in Pie Town, New Mexico, in June 1998, when I accepted an invitation to speak at a meeting organized by three local women who despaired over the social and economic cost that constant conflict between ranchers and environmentalists had brought to their communities. Jim and Joy despaired too, but for a different reason—the Williams Ranch was in trouble. In 1995, the Forest Service reviewed the Williams's grazing allotment and decided to cut the number of permitted cattle they could run on the forest. It was the first time the permit had been cut in Jim's lifetime, who was then in his early fifties. Moreover, it had never been cut during the lifetime of his father, Frank, who had assembled the ranch back in the 1940s. The issue of contention was the condition of the land, which the Forest Service insisted was being grazed too hard by Jim's cattle.

It was a common story at the time—with a common outcome. Angered by what he thought was the Forest Service's intransigence, Jim joined a class-action lawsuit with other ranchers against the agency. He also closely tracked another court case, this one brought by environmentalists upset at the government over cattle grazing on public land. "I thought the only answer was to fight," Jim told me later. "Well, we lost both of those cases, and so I thought that was pretty much the end of everything."

Financially struggling, and with their up-and-down relationship with the Forest Service at an all-time low, the Williams family, the last full-time ranchers in the Quemado area, began to seriously contemplate the one option that remained: to accept the offer of a subdivider to buy their substantial private land. Unwilling to take this option just yet, however, Jim raised his hand at the end of the meeting in Pie Town and invited the Quivira Coalition for a tour of his ranch, which we organized two months later. Liking what he heard us say about land health, progressive ranch management, and collaboration, Jim invited us back for further discussions. He also ordered the Catron County manager, who was on the tour, off his land when he tried to talk Jim out of cooperating.

Working with John Pierson, the Forest Service range conservationist, and range consultant Kirk Gadzia, Jim set new goals for the ranch and began to sketch out a new plan of cattle management. Using existing fences and natural boundaries, they divided the ranch into smaller pastures and planned rapid moves of cattle through them. Jim also agreed to graze his Largo Creek pasture in the winter months instead of late spring, as he had traditionally done. Jim and Joy also agreed to let Hawks Aloft, a nonprofit group hired by the Quivira Coalition, do bird monitoring on Largo Creek on their private land (to document the creek's ecological improvement). Everything went well—the grazing rotation worked, the land improved, and communication and trust between Jim and the Forest Service was restored. Jim even joined in on the bird surveys.

"I got a real kick out of looking for ferruginous hawks on my place," Jim told me, referring to an elusive and sensitive species of concern. That's probably not something a Catron County rancher would have said in the late 1990s.

An important fruit of this of trust building blossomed in 2001, when Jim and Joy opened their private land to the Quivira Coalition for a riparian restoration project along Largo Creek.

Not only was the creek in need of doctoring, but the ranch met an important precondition for Bill in any restoration project that he undertook—the livestock grazing had to be under control. There is no point to armoring a stream bank with riparian vegetation if the cattle come in and eat it all to the nub. As a consequence, Bill Zeedyk avoids working with landowners who overgraze, thereby creating a very important link between the New Ranch and riparian restoration. One reinforces the other—good cattle management helps the grass grow along the creek, and the riparian restoration can increase the amount of forage available for animals.

All of this involved a steep learning curve for me, but nothing quite prepared me for what happened when we turned our attention to a side tributary of Largo. It was called "Loco" for a reason—it was crazy to look at. Parts of it were so deeply entrenched that the walls rose above my head as I walked up it. According to Jim, it wasn't even a creek—it was an old wagon road that had eroded so badly over time that it intercepted the watercourse, redirecting it entirely. And it was eroding so badly that with each major cloudburst, its bottom could drop by a foot or more, kicking huge amounts of sediment into Largo Creek—which is not a good thing. In fact, it is precisely the sort of environmental trouble that has agencies like the EPA (which administers the Clean Water Act) worried across the region.

It wasn't the sediment that got my attention—it was something farther up Loco Creek.

Walking up the drainage one day, as crews were placing erosion-control structures farther down under Bill's direction, I came to the boundary between Jim's ranch and the national forest. Stretched across the creek and ten feet above my head was an old barbed-wire fence, complete with fence posts. I knew from a conversation with Jim that the Forest Service had built the fence in 1935. And the fence posts rested on the ground. In other words, a *huge* amount of erosion had taken place here in less than seventy years.

My God, I thought. I stared up at the fence for a while longer. Then I took a photo.

Later I asked a man who works for the Natural Resources Conservation Service (the old Soil Service), which is a branch of the U.S. Department of Agriculture that works with private landowners, how much of the rest of New Mexico existed in a degraded condition similar to what I saw in Loco.

"Most of it," he replied.

Enter Bill Zeedyk. How Bill came to his restoration career says as much about his generation, and how far it has come over the decades, as it does about him. Born in New Jersey to school-teacher parents in 1935, in what was then a rural area, he attended the University of New Hampshire, where he majored in forestry, having decided at the tender age of fourteen that he wanted to be a forester. He liked to hunt, fish, and trap—in fact, he paid for his first year at college by trapping muskrats. This led to his interest in habitat management—because he wanted to trap more muskrats.

"Trapping taught me how to observe wildlife and encouraged a sensitivity to habitat needs. It taught me how to read a landscape."

Despite his burgeoning respect for nature, however, Bill grew up in an era when humans assumed they knew best. "We were always looking for a better tool to control nature," he recalled. "That changed with Earth Day, when we began to see that there are consequences to all that we do. Up until then, we rarely took responsibility for our actions."

This included his employer. Bill joined the Forest Service right out of college in 1962, becoming the first biologist on the Daniel Boone (then Cumberland) National Forest in the mid-Appalachian Mountains. He believed firmly in the wisdom of multiple uses on public lands (and still does) because of its inclusiveness.

"Everyone stood to gain something from the common management of our forests, and this made the public lands system strong," he said. "Unfortunately, today the interests are splintered, and the support for public lands has eroded to the point where I believe their future may be in doubt. There is no longer the bond of common ownership that protected the integrity of the National Forest system."

As he rose through the ranks, he remained focused on the needs of wildlife. While in Washington, D.C., in the early 1970s, he helped draft the first policies for the Forest Service in implementing the Endangered Species Act. He was also on the front lines of the development of riparian management rules within the Forest Service. It didn't make him very popular. "No one valued riparian health back in the 1970s," he recalled. "One forest supervisor told me to my face to get lost. He said there were no riparian areas on his forest. It was all about timber and cows."

The unofficial attitude toward wildlife wasn't much better. There were few biologists employed by the agency, and the ones there were got caught up in intense turf battles with state wildlife agencies and the United States Fish and Wildlife Service. "The old thinking was get the range right [i.e., grazing management], then the wildlife will be okay," he said. "In the old days, 'wildlife' meant deer and elk, not much else."

Things began to change, however, mostly as a result of intense pressure from an environmental movement that was flush with victory at the time. Standards for what constituted healthy ecosystems rose, especially for riparian areas, but so did conflict among various interested parties, and stress among federal employees. In the late 1980s, as the chief wildlife biologist in the Southwest region, Bill got caught in the crossfire between activists on both sides of the endangered species issue. He tried for a while to walk a middle ground but soon exhausted himself from the constant friction.

By 1990, Bill was done arguing. He retired and tried to relax. But a personal tragedy and an enduring desire to make things better pushed Bill into his new career a few years later. Upon completing a series of classes with hydrologist and restoration pioneer Dave Rosgen, who Bill credits with organizing his own ideas, Bill was asked by Tom Morris of the Navajo Environmental Protection Agency to take a look at a serious erosion problem that endangered the western edge of Hubbell Trading Post, near Ganado, Arizona. The straightened, rapidly eroding creek, which carried tons of sediment from poorly managed lands upstream from the park, needed immediate attention. The park and the EPA were willing to give Bill's unconventional idea a try, and in the process, it became the first project where Bill could try "Induced Meandering" on a practical scale.

The creek responded quickly, sufficiently remeandering itself in a few short years for the park's structures to be considered safe from further erosion.

After his start at Hubbell, Bill fiddled with all of his ideas for riparian restoration over the next few years, only giving them a proper working out while on a consulting job in Mexico. Back in the States, Bill's ideas were greeted with a mixture of skepticism and outright resistance, especially by regulating agencies. Over time, however, as Bill's work proved itself where it mattered, on the back forty, the skepticism faded away. Today, many years after his retirement from the Forest Service, Bill has never been busier. He is, in fact, booked. He has worked across the Southwest, and beyond, and many of his ideas and techniques have been picked up by a new generation of restoration specialists. Of all the indicators employed to monitor the success of his work, this may be the most telling.

▪

TAPED TO MY computer is a postcard I found in a local coffee store. It depicts an ill-looking planet Earth, with its tongue hanging out, imprinted with the message: "The world could be in better shape." Surrounding this image are words: *renew*, *heal*, *reaffirm*, *nurture*, *rekindle*, *revitalize*, *repair*, *revive*, *mend*, *soothe*, *rebuild*, *fix*, *regenerate*, and *reinvigorate*.

I've thought a lot about those words over the years as the Quivira Coalition worked with progressive ranchers to restore land to health and embarked on a series of substantial riparian projects under the direction of Bill Zeedyk. They are words of advancement and action—positive, progressive, healing action. By contrast, much of the vocabulary I learned as an environmental activist focused on defense or safekeeping—*save*, *preserve*, *roll back*, *stop*, *protect*, *prohibit*, *enforce*. This vocabulary is still needed as we head deeper into the century, but I've come to believe that it is more the language of *healing* that gives people meaningful direction and hope.

People respond to restoration work because it involves us in a giving rather than a taking—a giving back to nature, an honoring, while we necessarily continue to take nature's bounty. We can't stop using nature—we need its air, its water, its food, its animals, its minerals, its beauty, and its inspiration for our well-being. We must take, but *how* we take, as well as what we do with what we take and what we leave behind, lies at the root of many of our environmental troubles. As we take, we can also give—and not just for the gesture's sake. Giving is becoming a requirement. The world not only *could be* in better shape—it *must be*, and soon, according to many experts and elders. The survival of the earth's biota (including us) requires that we renew, heal, reaffirm, nurture, rekindle, revitalize, repair, revive, mend, soothe, rebuild, fix, regenerate, and reinvigorate the planet's natural heritage.

But there is another reason why I like these words, something beyond the practical and the doctoring. They are words of redemption. It has to do with the way we treat each other, the damage we do to relationships with one another, with nature, as well as to the trouble we cause natural processes. We rarely seek redemption in our daily lives, mostly because we live in an age and a society that has almost completely buffered us from the consequences of our actions. We eat, drink, travel, and consume without retribution. Furthermore, a wide variety of cultural agents—including the TV, the grocery store, the automobile, the city—tell me I don't need to worry about giving back. Their message is clear: Keep taking. All is well.

All is not well, of course. But I knew that going in as an environmental activist. What I learned over time, however, is that we *can* make things better, not by shielding a special place from all this taking going on but by *giving,* and in so doing, try, even with small gestures, to redeem ourselves. In other words, the restoration of health—to creeks, grasslands, ourselves—is a kind of moral exercise. I'm not sure that Bill Zeedyk or any number of ranchers I know look at it quite in those terms, of course.

But I do.

TWENTY

CONSERVATION IN THE AGE OF CONSEQUENCES

(2007)

"We are not walking a prepared path."

—**WENDELL BERRY,**
at the Quivira Coalition's Sixth Annual Conference,
in response to a question about the difficulties that lie ahead

IN JUNE 2006, forty-nine heifers were delivered to the Quivira Coalition's ranch on the thirty-six-thousand-acre Valle Grande allotment on the Santa Fe National Forest atop Rowe Mesa, southwest of Santa Fe, New Mexico, and just like that, a bunch of conservationists became ranchers.

They were the first installment of what would become a 124-head herd of heifers, plus three Corriente bulls, all under our Valle Grande brand, and all under our management.

This was an intriguing turn of events for the staff and board of the Quivira Coalition, a nonprofit whose original mission was to create common ground between ranchers and environmentalists. It was also a surprising twist for me personally. If ten years ago you had told this former Sierra Club activist that I would be in the livestock business, selling local beef to Santa Fe residents, I simply would not have believed you. But here I am—a dues-paying member of the New Mexico Cattlegrowers' Association.

Maybe it wasn't such a stretch. After ten years of encouraging ranchers to act more like conservationists, it suddenly seemed logical that we, as a conservation organization, begin to act more like ranchers. It wasn't just a matter of walking the walk either—the harder we looked, the more conservation opportunities we saw running the ranch *as a ranch*.

In fact, when discussing this turn of events in my lectures around the region today, I state simply that the Quivira Coalition is "a conservation organization that manages livestock for land health and prosperity."

I thought all this was something new under the sun. But what exactly?

To gain perspective, I reread Charles Wilkinson's classic study *Crossing the Next Meridian: Land, Water, and the Future West*, published in 1994, which I knew to be a thoughtful analysis of late-twentieth-century conservation. In it, I read that the major challenge for activists nearly twenty years ago was grappling with the legacy of the "lords of yesterday"—the laws, customs, and policies created in the wake of the West's vigorous frontier era.

These "lords" include the 1872 Mining Act, which encouraged a fire sale of public lands to mining interests; the 1902 Newlands Act, which inaugurated an era of frenzied dam building; the implementation of the "Western Range" concept in 1905 (and the follow-up 1932 Taylor Grazing Act), which institutionalized livestock interests on public land; as well as various timber, homestead, and water laws and regulations.

By the 1980s, Wilkinson wrote, these "lords" were out of kilter with the urban public's burgeoning interest in outdoor recreation and the protection of natural resources, resulting in a great deal of conflict with rural residents across the region. From the "timber wars" of the Northwest, the "grazing wars" of the Southwest, the "wolf wars" of the northern Rockies, and the clashes over endangered species nearly everywhere, the struggle between the "old" West and the "new" kicked into high gear.

For nonprofit conservation organizations of the era, their mission was straightforward: fight *for* wilderness areas and national parks and *against* the "lords of yesterday." On the economic side of things, these groups touted the tonic of increased recreation and tourism, whose mostly unquestioned benefits were blossoming at the time of the publication of Wilkinson's book.

This mission caused two types of conservation organizations to bloom. The first was the advocacy-based organization, sometimes called the "watchdog" model, whose mission was to challenge wrongdoers and protect environmental values, principally on public land. Often this meant fighting the federal government—and by extension, miners, loggers, and ranchers—in court as well as in the court of public opinion.

Concurrently, another type of conservation nonprofit formed in response to threats posed to the natural assets of private land. The modus operandi of these groups was preservation by purchase—buy it, save it—sometimes called the "trust" model, though they also leveraged land transfers to federal and state agencies.

Together, the "fight it, buy it" counterpunch to the "lords of yesterday" netted significant results, including a raft of important federal laws, which unquestionably improved the quality of life for wildlife and humans alike.

Fast-forward to 2007, however, and both the problems and the cures for the American West as identified in *Crossing the Next Meridian* seemed out of date. This is not Wilkinson's fault; rather, it is a sign of how much things have changed. For example, Wilkinson makes little or no reference to global climate change, restoration, collaboration, the rise of watershed groups, the expansion of local food markets, or the dynamic energy of agroecology, though he does identify the outlines of the progressive ranching movement. Similarly, there is little mention of the downside to an amenity-based economy, including the damage widespread suburban and exurban sprawl would soon do to communities of people and wildlife.

He does talk about sustainability—much in the news these days—and concludes his book with a call for "sustainable development" in the West, though the main mechanism he proposes for achieving it is the planning and zoning toolbox. Presciently, he speculates that the journey to a sustainable West will be a long one.

My frustration with the divisiveness of the "fight it, buy it" models led me to cofound the Quivira Coalition in 1997 with a rancher and a fellow conservationist. One of our original goals was peacemaking, exemplified by our tagline at the time: "Sharing Common-sense Solutions to the Rangeland Conflict." But ten years later, the question on my mind was this: where did the Quivira Coalition fit in exactly? We weren't a watershed group, nor did we labor to achieve consensus among stakeholders or mediate conflicts over natural resource use. Instead, we worked regionally, aimed our efforts at "eager learners," and promoted a land stewardship toolbox that focused on land health. Later, we moved into land restoration projects. Eventually, we became ranchers.

It felt like we were walking a new path, but to where?

Although no one knows what the decades ahead will bring precisely, there are enough indicators of change to say with confidence that the challenges will be varied and daunting. Some are already here, including widespread land fragmentation, the expansion of destructive industrial agricultural practices, the compounding effects of population pressures, a burgeoning "over-recreation" of our public lands, a dissolving bond between nature and young people, and the effect of all of the above on biodiversity.

These are all elements of what I call the Age of Consequences. I like to think of this age as a hurricane that has been building slowly over open water and is now approaching shore. We can already feel its winds. We don't know precisely where the bulk of the hurricane will make landfall or how strong its winds will be ultimately, but we do know that it will strike and that its destructive power will be awesome.

A strenuous effort must be made to lower the wind speed of this hurricane as much as possible—such as reducing the amount of greenhouse gases entering the atmosphere or preserving biologically rich natural areas from industrial development—which are great roles for the "fight it, buy it" school of conservation. At the same time, we must acknowledge the inevitability of the hurricane's landfall. That means a simultaneous effort must be made to increase ecological and economic resilience among landowners, organizations, and communities so that they can weather the coming storm of change. It's also what the Quivira Coalition has been trying to accomplish over the past decade, though we didn't think of it in those terms at the time.

We do now.

Resilience is the ability to recover from or adjust easily to misfortune or change. In ecology, it refers to the capacity of plant and animal populations to resist or recover from disruption and degradation caused by fire, flood, drought, insect infestation, or other disturbance. Resilience also describes a community's ability to adjust to incremental change, such as a slow shift in rainfall patterns or a rise in temperatures.

The word also has a social dimension. Ranching, for instance, is the epitome of resilience, having endured a century of cyclical drought and low cattle prices, as well as a host of modern challenges. Of course, some ranches were not strong enough to ride out the storm, succumbing to sprawl, bankruptcy, or the loss of the next generation, but many endure and are finding ways to keep their ranches going.

For those of us who live in cities, there is a lot to think about in resilience. Take food, for instance. If there were a major disruption in our food supply, what would we do? Where would next week's meal come from? Are there enough farms and ranches in the area to feed all of us? Do we have enough resilience to weather an energy crisis or a water shortage?

Building resilience means many things, but for the purposes of conservation work in the future, I believe there are three main areas of focus:

- *Reversing Ecosystem Service Decline.* In 2005, the United Nations published its *Millennium Ecosystem Assessment*, a global evaluation of the ecosystem services on which human well-being vitally depends. These services include food, fresh water, wood, fiber, fuel, and biodiversity; climate, flood, pests, and disease regulation; nutrient cycling, soil stability, biotic integrity, watershed function, and photosynthesis; and spiritual, educational, recreational, and aesthetic experiences.

The basic conclusion of the assessment is this: globally, ecosystem services are in decline, and as they go, so will human well-being. And as human well-being degrades (and it's already started in many places around the globe), traditional conservation concerns, such as wilderness protection, parks, and recreational experiences, will fall in priority. That's because as conservation strategies, they'll be less and less effective, as basic human needs, such as meeting food and energy requirements, rise in importance.

The assessment's authors make much the same point. To reverse the decline in ecosystem services, they encourage *active adaptive management*—experimentation and monitoring with new management methods—to maintain "diversity, functional groups, and trophic levels while mitigating chronic stress [in order to] increase the supply and resilience of ecosystem services and decrease the risk of large losses of ecosystem services."

In other words, conservation will shift from protection and preservation to restoration and management—from saving land to working it properly.

- *Creating Sustainable Prosperity.* Ecosystem services have declined partly because their conservation value is not seen to be in the economic self-interest of important portions of society. As a result, conservation, including the restoration and maintenance of natural systems, became primarily a subsidized activity, accomplishing its goals principally (1) by direct or indirect governmental funding, (2) as an indirect product of commercial agricultural activity, (3) by philanthropy, or (4) by some combination of each.

Conservation remains subsidized for a variety of reasons, including its high cost. Another reason is a well-founded concern about the role uninhibited market forces play in the overexploitation of natural resources—a role that has contributed widely to ecosystem service decline around the planet.

But can conservation pay for itself? If it cannot, at least at some significant level, then the objective of reversing the decline of the ecosystem services on which human well-being depends might be impossible. That's because more than a century of conservation work has demonstrated the limitations of subsidized incentives (case in point: the current condition of the planet). Additionally, the scale of the conservation job continues to grow, especially as ecosystems decline, which means the cost of restoration will grow as well.

But even if conservation can be profitable, can it be *prosperous?* For many family-scale progressive ranchers, the answer is "yes." They've done it by working on the original solar power, as grass farmers. Many have been profitable and sustainable simultaneously, and often for the same reason, thus prospering in multiple ways, and not just economically.

- *Relocalization.* The inevitability of rising energy costs means more and more of our daily lives, from food production to where we work and play, will be lived closer to home at local and regional scales. This won't be by choice, as it is currently, but by necessity. The key is to look at relocalization as an opportunity, not just a challenge. It can be a form of rediscovery—learning about our roots, about community, neighbors, gardens, and doing with less in general. One could even look at relocalization entrepreneurially—those individuals and organizations that get into the game early, by providing relocalized goods and services, will stand a very good chance at a profitable living.

Working landscapes will become critical again. So will local farmers and ranchers. This means figuring out how to keep the current generation of farmers and ranchers on the land, as well as encourage the next generation to stay, come back, or give agriculture a try. Producing local food and energy from working landscapes will also require healthy land and best management practices that work within nature's model. While the toolbox of progressive stewardship is now well developed, a great deal of our land is still in poor condition, requiring restoration and remediation.

Paychecks are among the simplest solution to these challenges.

Lastly, I believe pressure will build on the federal land agencies to adopt comanagement principles with private organizations on public land. Bureaucratic gridlock combined with a persistent inability or unwillingness to innovate on the part of the land agencies means that partnerships with private entities, including a new generation of grazing permittees, are the only viable means to meet ecosystems service and relocalization challenges on public land.

All of this work involves creating a "new path"—to paraphrase Wendell Berry—since many of the challenges that it addresses are novel. The "fight it, buy it" models of conservation, which have an important role to play in slowing the hurricane down as much as possible, alone are no match for the big job of resilience.

The challenges of the Age of Consequences require a new type of conservation organization. In fact, I'll postulate that reversing the decline in ecosystem services on which human well-being depends will ultimately prove to be the primary mission of conservation in the twenty-first century.

Reversing ecosystem service decline, however, requires adopting a simple but radical new philosophy: that all natural landscapes must now be actively managed. Some may need more management than others depending on the level of resilience required, but under the global effect of climate change, we can no longer turn our backs on our responsibilities, no matter how big or small.

For ranchers and conservationists alike, this means doing things differently. We can get started by restoring land to health, by producing food locally, by sharing information and resources, by working together, and by looking and learning.

One stepping-stone at a time.

TWENTY-ONE

BIG THINGS IN SMALL PLACES

(2008)

SANDIA PUEBLO

NORTH OF ALBUQUERQUE, NEW MEXICO

I LOVE THE words range professionals use to describe the elements of ecosystem function: *integrity*, *diversity*, *resistance*, *thresholds*, *transitions*, *recovery*, and so forth. That's where I found *resilience*. It describes the ability of a community to recover from change or misfortune—how it handles surprise, in other words. And nature is full of surprises, as we all know. How a community of plants or animals bounces back from an unexpected flood, drought, disease outbreak, fire, hurricane, or other perturbation depends largely on its health—its ability to resist degradation while the event is occurring and its capacity to recover once the surprise has ended.

But what does resilience actually look like? To find out, I decided to visit Sam Montoya again.

Sam is a tribal elder of Sandia Pueblo, a Native American reservation located a few miles north of Albuquerque. What he had accomplished on his very small farm, I recalled, was not only impressive, but very possibly quite resilient. So I wanted to see how things were going. When I first met Sam six years ago, I was astonished to see 220 head of cattle grazing on the ninety-three acres of his little farm.

That's right: *220 cattle on only ninety-three acres of land.* In the arid Southwest, that many cattle typically need a bigger spread—a *much* bigger spread. For example, the Quivira Coalition runs between two hundred and three hundred cattle on a public allotment that is thirty-six thousand acres large. As you might suspect, the difference is water—Sam's little farm is irrigated, but that only makes his story even more intriguing.

Here is what I learned from Sam during my first visit: after retiring from a career with the Bureau of Indian Affairs, Sam decided that he wanted to return to his agricultural roots. Upon receiving permission from the tribe to rehabilitate ninety-three depleted acres of a former sod farm, located a short distance from his home, he laser-leveled the land, built a central watering source, planted orchard and other grasses, and then divided the ground into thirty-three paddocks—three acres each—with electric fencing. Then he turned the water on. When the last dairy in the area shut down due to subdivision pressures, Sam scored a natural and economical supply of fertilizer. When the grass grew lush, he turned the cattle out.

The animals graze as a single herd in one paddock for one day only. When the twenty-four-hour period is up, Sam drives over from his house, lowers a gate in the electric fence, watches as the cattle drift into the adjacent paddock, secures the fence when the move is complete, and goes to work. The entire process takes less than half an hour, meeting Sam's requirement that he "not work too hard" in his retirement. The rotation through the paddocks takes a little more than a month, by which time the irrigated grass is ready for another harvesting. And he repeats this cycle all year round.

"I'm trying to mimic what the bison did," said Sam. "They kept moving all the time. You, me, the land—everything needs a break. But you shouldn't sit on the sofa all week. Too much rest is as bad as too much work. It's all about balance."

Pursuing that balance, Sam didn't use pesticides, herbicides, or other chemicals. And other than the delivery and pickup of the cattle, Sam's operation required *no* fossil-fuel-dependent machinery—a fact that pleased the economically minded farmer.

"I don't want anything that rusts, rots, or depreciates," said Sam, grinning. "Plus, I feel good that I'm not polluting the air."

Today, he could add something new to that statement: he's not contributing to global warming either. That's because his operation worked on the original solar power: photosynthesis. In fact, Sam called himself a "grass farmer"—which meant he considered grass to be his principle product, not beef. The cattle are his lawn mowers.

Perhaps as important as anything else, Sam was making money. Profits from the sale of cattle—Sam is a studious observer of business cycles in the livestock industry—allowed him to quickly pay back the loan he took out to get the farm started. After only a few years, he operated in the black—undoubtedly due to his very low costs.

In sum, Sam's little farm seemed to be a perfect illustration of resilience. He operated almost entirely off the industrial grid, producing healthy animals raised on grass managed in a way that mimicked nature's model of herbivory. He recycled everything and wasted nothing. Short of a natural catastrophe, Sam's farm would probably survive whatever surprise the world threw at it.

Or could it? That's why I returned for a visit—how had Sam and his little farm held up over the years? Was he actually *being* resilient?

I knew the quick answer was "yes." That's because every time I drove from Santa Fe to Albuquerque on the freeway, I could see Sam's cattle grazing on their patch of heavenly green near the Rio Grande River. It's quite an anachronistic vision too—a little farm wedged between the busy interstate to the east, smoggy Albuquerque to the south, and rapidly growing Rio Rancho, home to a major computer chip–manufacturing complex, to the west. Perhaps I should also mention the large casino at the border between Albuquerque and Sandia Pueblo, operated by the tribe, which added to the time-out-of-place feel of the little farm.

But Sam's little ranch was no more a mirage today than it was six years ago. The lush grass is real and the fat cows are real, as are Sam's profits, he reported.

We met near the tall cottonwood tree that dominates the farm. A handsome man with a distinguished amount of gray in his otherwise black hair and a low-key but infectious smile, Sam looked relaxed. He still wasn't working too hard, he said, though most of his time was taken up directing a project for the tribe to preserve his native Tiwa language.

As for the farm, the only thing that had changed was his decision to sell his cattle, at the top of the cattle cycle, some years ago. Now he grazes cattle, for a fee, for other pueblos and individuals. Otherwise, everything was working smoothly, he said.

We walked out to the cattle in a paddock, dodging numerous manure piles in the green grass. The animals watched us docilely. Not far away, a large flock of Canada geese grazed peacefully. "It works pretty well," said Sam of the farm, stopping to rub the head of his favorite bovine. "I guess you could call it resilient. It's been pretty good to me. And I know it's been good for the land. Sometimes too good—I have trouble keeping ahead of the grass sometimes." At least he doesn't have to worry about the market anymore. By feeding other people's cattle for a certain price per head per day, he doesn't have to worry about fluctuations in the market—guaranteeing him a good price regardless of the price of cattle.

Sam confessed to only two disappointments with his work, both of which are interconnected. First, despite his obvious success agriculturally and financially, no other "grass farm" has been established on the reservation since he began his endeavor. None of his peers seem particularly interested in his farm—a fact, Sam says, that is directly related to the success of the nearby casino. But even his farming neighbors aren't curious. One continues to work with big machinery—and burn fossil fuels.

The second disappointment hits closer to home, I think. No one in the pueblo can get members of the next generation very interested in agriculture. Sam sees a parallel with his work to preserve his native language. Kids today have too many competing interests, he said, including the lure of the expanding digital universe. "Some come out to see what's going on here," he told me, "but no one wants to go into agriculture. I don't blame them. After all, the tribe will pay for school so they can become doctors and lawyers."

We wandered back across the paddock, examining the fine condition of his cattle. The grass looked pretty good too. Of course, the irrigation helps, but that's the point: resiliency isn't abstract. It requires soil, water, air, and sunlight to thrive. And in the arid Southwest, water in particular gives the land fertility—its regenerative capacity to grow, die, and grow again. But not too much water. It's all about balance, as Sam observed. Too much of a good thing can be bad for you in the long run. That's why Sam manages his water carefully, applying neither too little nor too much, but just enough to stay resilient.

We reached our trucks near the cottonwood tree. We talked about lessons learned, about money, taxes, the cycles of nature, and the marketplace. Resiliency is a complicated word, we decided. It can't be accomplished alone—it needs to be part of a community effort. By getting off the industrial grid, Sam made his farm resistant to surprise—he made it sustainable, in other words—but only to a point. He is still working by himself, which raises an important question: what happens when Sam *really* decides that he doesn't want to work too hard?

MINE TAILINGS
GLOBE, ARIZONA

PULLING UP AT a stoplight in Globe, Arizona, during a spring break journey with my family, I casually glanced at the hill to my right—where I saw something moving. Craning my neck, I peered through the windshield and saw maybe thirty cattle grazing peacefully on the slope, bunched together in a tight herd.

The hill was actually a huge pile of mine tailings—where the waste rock from decades of open-pit copper mining was hauled and dumped. From a distance, it looked like a giant steep-sided ziggurat—an ancient Mesopotamian edifice that rises in levels from a massive base—only with cattle grazing on its side! This might seem incongruous to many people, correctly, but I knew what was going on.

The main problem with mining, of course, is not the mineral extracted but the waste left behind. Whether it is a lone prospector hauling ore out of a shaft or a multinational corporation moving mountains, mining is messy, to say the least. Since the rock is excavated far below the surface, it is essentially sterile—colorful, perhaps, but lifeless. When piled high, it quickly erodes, especially after a torrential summer thunderstorm.

Unregulated, poorly designed, and poorly executed mining has caused a litany of environmental damage around the world. I won't go into their sins here, which have been well documented, other than to say that there's nothing redeeming about an open-pit mine other than its awesome scale. And the grass. This was no sterile pile of rock any longer—it was covered with vegetation. For confirmation, I turned the truck around and drove to the eastern side of the ziggurat, where, as I expected, I saw grass—lots of it. The cows had worked, over time, around the tailing, and apparently it had rained in the interim. We had tried something similar years ago on a mine tailing in New Mexico, albeit on a much smaller scale. Our goal had been to grow grass—life—on largely lifeless soil. And for a while, it worked.

At this point, you may be wondering: Cattle grazing on a mine tailing? What is he talking about?

The quick answer is that it's called a "poop-n-stomp"—a name I made up to describe our little mine reclamation project. Not only did it convey our employment of cattle as agents of restoration, but it was also a literal description of the process.

In early 1999, I received a phone call from an EPA administrator in Dallas, Texas, who said they had some extra money in a Clean Water Act account and asked if I might be interested in conducting a restoration project with it. He knew that our little start-up nonprofit, which focused on the ecological benefits of good livestock management, was eager to implement demonstration projects. When he specifically suggested mining, whose eroding tailings are a perpetual source of headaches for his agency, I said, "You bet."

That's because I knew who to call.

I had recently met Terry Wheeler, a feisty and outspoken rancher from Globe, who had successfully pioneered a mine-reclamation strategy that used only livestock, hay, grass seed, electric fencing, a portable water source, one or two humans, and not much else. His idea was as simple as it was brilliant: build a small paddock on a patch of eroded slope, spread the grass seed and hay across the ground, turn out the cows for a few days, and watch as they press the seed into the ground with their hooves while eating the hay. Add the bodily functions of the livestock, rain, and presto! Green grass.

It was no different, Terry liked to observe, than the instructions on the back of a packet of seeds that you buy to plant in your garden: Press seed firmly into soil. Just add water. The only variables in this case were the hay (a carbon source), the nature of the fertilizing process, and the seven-hundred-pound animals who did most of the work on a forty-degree sterile slope.

As Terry tells the story, when he first approached the owners of a copper mine in Globe with his idea, they were both curious and skeptical. Curious because mine reclamation is a big challenge for many companies—it's expensive, time consuming, difficult, a source of conflict with regulating agencies, and prone to failure. They were skeptical because no one had proposed using animals to do this work.

Many traditional reclamation strategies involve costly combinations of water pipelines, mechanical sprayers, chemical fertilizers, diesel-powered machines, and human labor. The goal is to stabilize the tailings so they won't erode into a nearby creek, and if the process is not designed properly, implemented correctly, or maintained adequately, then all that work and money is often literally washed away in a few years. So when Terry told the mine owners that he could reclaim one of their massive tailings for less money and with better results, using an organic process instead, he got their attention. Their skepticism kicked in when he said he would do the work with cattle.

"One mining executive," Terry told me, "liked to joke that they should line up BBQ grills at the bottom of the slope for all the cattle that would come tumbling down."

The cattle didn't, of course, come tumbling down. They did just fine, pooping and stomping their way back and forth across the tailing, pressing the grass seeds firmly into the soil with their hooves. When the rain came and the grass grew, Terry said, the jokes stopped.

When I hired Terry to do our little project—a twenty-acre patch of eroding soil on an abandoned copper mine near Cuba, New Mexico—using the EPA funds, I had a different objective in mind. I was intrigued by the possibility of using cattle in the service of environmental restoration. In fact, he didn't think of his cattle as cattle. Instead, he called them FLOSBies—Four-Legged Organic Soil Builders.

And that's exactly what they did for us over the course of two summers on that New Mexico copper mine—build soil and grow grass. But they did more than that. Though we didn't talk about it in these terms at the time, what Terry's FLOSBies were doing was building resilience—and not just in the soil. For a society fixated on technical and petroleum-based solutions to every problem—many of which are proving to not be very sustainable—it was inspirational to discover an organic alternative that could be effective, redeemable, and profitable! I saw all of the above on our little reclamation project in New Mexico.

Unfortunately, for all of its achievement, our little restoration project outside Cuba eventually turned into a pumpkin, teaching me, in the process, a lesson about a deeper definition of success—and, ultimately, resilience. Ecologically, our reclamation results were great, at least initially. Over two summers, Terry's herd of FLOSBies poop-n-stomped those twenty acres back to life. Winter snows and spring rains caused the slopes to grow a great deal of grass. Soil stabilized, gullies healed, rain soaked in instead of running off, and the ground turned green during the summer. Various agencies, including the EPA, were pleased.

Returning to the mine two years later, however, I was surprised to discover that nearly all the grass was gone. At first I suspected the ongoing drought, but as I walked through the project site, I came across the real culprit: trespass cattle. Unfenced, the grass had disappeared into the bellies of local herbivores.

I reflected on this unexpected turn of events. Although we had the cooperation of the private landowner, a local rancher, I realized that we had failed to engage him meaningfully in the mine project. He gave us permission to do the work, but he gave us little else. He never became a real partner. It was our project, not his—or the community's. When we left, cows appeared. Nobody was to blame, but it taught me a lesson about local buy-in. Innovation can't be imposed from outside.

Three years after a wet winter, I returned to our former restoration project and was pleasantly surprised to see grass. I parked my truck, grabbed the camera, and climbed the steep slope of the tailing. Apparently there was enough seed and straw still in the soil to get grass growing again. It was a pretty sight to see.

It was resilience in action.

All of this came rushing back to me during our brief stop in Globe, observing what were very likely Terry's cattle at work. I was happy to see that Terry's unorthodox idea was still alive on another waste pile, his FLOSBies still creating life. I snapped a photo. That's what I like about resilience—the only thing that matters in the long run is what sticks around.

TWENTY-TWO

NO ORDINARY BURGER

(2009)

CAN A HAMBURGER save the family ranch in the twenty-first century?

If you're the Diablo Burger, a bite-sized eatery located in the busy old-town heart of Flagstaff, Arizona, that serves up natural, fresh, trendy, and tasty hamburgers supplied by two local ranches, the answer is: possibly. *Hopefully*. The restaurant also features Belgian-style fries; hormone-free whole milk milkshakes; herbs, onions, and tomatoes from local farms; bread and cookies from a bakery in Phoenix; citrus from McClendon's Select farm in Peoria, Arizona; and ice cream from the Straus Family Creamery, located north of San Francisco, California.

This is hopeful news because the entrepreneurial, privately owned restaurant is an example of an effort by the Diablo Trust, a pioneering collaborative nonprofit, to encourage diversified business opportunities for ranches in the area. It's no ordinary burger, in other words, not simply because of the fancy fries or trendy music, but also because of what it symbolizes: the rise of a local economy that serves the cause of ranchers, city residents, and conservationists alike.

The Diablo Trust believes that strong family ranches maintain a healthy rural economy and culture while protecting open space from development. The question is: what does this mean in the early twenty-first century? For ranchers, it means innovating their age-old business model in order to develop new markets for their products. For city residents, it means participating in a local economy, especially as farmers' markets and other forms of sustainable agriculture expand. For conservationists, especially those who worry about the loss of open space to subdivisions, it means rethinking the way private land traditionally gets protected in the West, including age-old prejudices about livestock. For each, it means keeping the "work" in working landscapes—which is good business for everyone.

But let's back up and put Flagstaff's devilish burger in a broader context.

Of the American West's approximately one million square miles (roughly a third of the nation as a whole), half is publicly owned as national forests and parks, military reservations, wildlife refuges, or by the Bureau of Land Management (BLM). The other half of the West—approximately the size of California, Oregon, Washington, Arizona, and Nevada combined—is privately owned or part of sovereign Native American nations. Furthermore, homesteaders in the late nineteenth and early twentieth centuries took the best land first, meaning the most productive, well-watered, and least snowy (lower elevation) parcels. Not coincidently, this privately-owned land is the site today of high concentrations of biodiversity, especially in riparian corridors and wetlands. According to some estimates, as much as 60 percent of endangered species in the West exists on private land, much of it owned by ranchers. For these reasons and more, private ranches are seen now as critical pieces in the conservation puzzle out West.

Unfortunately, it is precisely this land that came into the crosshairs of developers in the early 1990s as the economy boomed and many urban refugees fled to the rural West. By 2005, the process of ranch and farm conversion to subdivisions reached an alarming rate of *one acre per hour*. This fact caused many of us in the conservation movement to realize that subdivisions were a greater threat to the region's biological diversity than the overgrazing crisis on public land that I had been repeatedly told by my peers was supposedly ruining the West. Instead, I learned that grass, when given enough rain, is quite resilient. The deleterious effects of a subdivision on the land, in contrast, were not so easily reversed. I also learned that ranches are resilient too, given the right economic and social conditions.

And yet, the typical response of conservation organizations to the open-space crisis was to buy a farm or ranch outright when it came on the market, at high cost, or facilitate the purchase of its development rights via a conservation easement. This strategy has been effective, but only up to a point, for two reasons: first, it requires a lot of money, which means conservationists will always be at a disadvantage to developers; and second, this "buy it" strategy often means the cessation of the land's agricultural productivity, resulting in a loss of community, history, culture, natural resources, and other benefits. This is why the effort to save ranch and farmland from development over the past two decades or so, while successful in some spots, has come up short in others, such as the Front Range of Colorado, for instance.

Fortunately, there is another way to protect private land—a way that ranchers, city residents, and conservationists can work together. From my experience, I believe the most economical and long-lasting way to protect privately owned open space in the American West from development is to keep productive ranches in business. It is far cheaper to help ranchers diversify income streams and create supportive collaborative relationships than it is to purchase their ranches on the open market or arrange for easements on their properties.

I call it the "Not 4 Sale" strategy. But implementing it requires cracking a difficult paradox: while many ranchers don't want to sell out to developers, many can't afford to stay in business either. Many landowners stay in ranching, I've observed, not *because* of the economic returns of commodity livestock production, but *in spite* of them. This is why ranching is sometimes described by academics as an "irrational" economic enterprise, for its dismal profit margins. This fact is supported by ranchers themselves who almost always list the social and cultural benefits of their way of life ahead of profit making. Still, ranchers have bills to pay like everyone else. Hanging a "Not 4 Sale" sign on the front gate of a ranch means finding a way to pay those bills, which has become more difficult in recent years.

The answer is to blend the needs of ranchers, city residents, and conservationists into a diverse suite of options that the keep the "work" in working landscapes. They include:

- *Increased Profitability*. Many ranchers have begun to diversify their income streams in an effort to remain profitable. Examples include (1) increased stocking rate as a result of progressive livestock management; (2) fees from hunting, fishing, camping, wildlife viewing, bed-and-breakfast services, dude ranching, and other amenity-based activities that attract urban visitors; (3) grants from foundations and agencies for a variety of ranch and watershed-based improvements, including the creation of local 501c3 organizations; (4) participation in local cooperatives that add value to ranch products; and (5) involvement in wind or solar energy projects, conservation projects (easements), or small-scale developments (a few home sites), that create additional revenue for the ranch operation.

- *Collaborative Networks*. Starting in the mid-1990s, landowners across the West began to see the strength in partnerships. Initially, most collaboration was defensive—pushing back against this or that threat—but over time, they evolved into proactive enterprises that brought economic opportunities to the region. They also spurred innovation—partners often have different skill sets, a new perspective, or access to resources unavailable to a single landowner. Also, friendship is critical to the political process and to policy reform.

- *Restoration*. The entrepreneurial opportunities for landowners to restore damaged or degraded land to health are growing rapidly. Examples include using livestock to control noxious weeds; using "controlled grazing" impacts (similar to controlled fires) to achieve desired ecological goals; conducting riparian and upland restoration work for water quality and wildlife habitat goals; tackling forest health concerns through thinning and other projects; repairing and upgrading low-standard ranch roads so they can restore natural hydrological cycles; and working collaboratively on watershed-scale initiatives to improve the overall health of the area, which increases productivity, which helps the bottom line.

- *Local Food Production*. There has been an explosion of interest in recent years among city residents in local, organic, natural, and food, which can mean increased social and economic profitability for ranchers. Grass-fed beef, for instance, can command 50 percent more per pound in price than commodity (feedlot) beef. Almost as important are the social and emotional benefits of getting into local food markets, including direct contact with customers, who often become advocates for the farm or ranch.

- *Other Ecosystem Services*. For centuries, well-managed farms and ranches have been delivering "ecosystem services" to cities, such as healthy topsoil, wildlife habitat, clean water, fuel sources, food, functioning wetlands, and buffers against floods and fires. It is only recently, however, that these services have come to be recognized, and therefore valued, as something worthy of protecting, restoring, and maintaining, especially as urban populations grow and pressure mounts on natural resources.

The story of the Diablo Trust is a good illustration of how family ranches are employing these strategies in order to stay intact during rapidly changing times. The story began in 1993, when the owners of the Flying M and the Bar T Bar ranches, located southeast of Flagstaff and comprising 426,000 acres of public and private land, decided to join forces and try a new idea: collaborative conservation.

Both ranches were struggling economically and emotionally. Despite adopting innovative range-management practices, including short-duration grazing, the ranches were forced to take stock reductions to alleviate what were perceived by state and federal agencies as conflicts between cattle and wildlife. The toll wore down the owners of the Flying M and the Bar T Bar. They contemplated selling out.

Instead, they formed the nonprofit Diablo Trust in order to enlist diverse community support for the ranches and assist with many of the nonranching challenges that confronted them on a daily basis. It was a big gamble. When over one hundred people attended the first meeting, including many members of agencies and some environmentalists, they knew they were onto something important. Committees were quickly established to focus on specific concerns, such as recreation and wildlife. A facilitator was hired to help the members of the trust reach consensus, a volunteer director was hired, and a mission formulated, which read: "The purpose of the Diablo Trust is to maintain ranches as long-term, economically viable enterprises managed in harmony with the natural environment and the broader community."

Did they succeed? Yes. The trust meets every second Friday of the month; has a variety of working groups; raises money for science, education, and monitoring projects; conducts community outreach programs, including an annual art-on-the-ranch day; publishes a regular newsletter; and strives to accomplish its vision through collaboration and innovation.

But perhaps the best measurement of success is this one: not one acre of private land on either ranch has been developed since the trust's founding. By working proactively with federal and state agencies, instead of reactively; by seeking partnerships with conservation groups, rather than assuming a defensive attitude all the time; and by reaching out to the public constructively, the trust helped the ranches stay in business. In other words, the "Not 4 Sale" signs on their gates were never taken down, thanks to success of the partnerships fostered by the trust and its programs. But is it enough to keep the ranches going until the twenty-second century?

This is where the tasty burger comes in.

The tiny restaurant that opened in early 2009 and has become a successful enterprise. The meat for its hamburgers is supplied by the Flying M and Bar T Bar ranches, which is part of the restaurant's pitch to its primary customers: residents, not tourists. Local food for local people. The restaurant takes only cash—in order to keep the money in the local economy.

Why local? Here's what the Diablo Burger menu said when I visited: "Because local food retains more nutrients; because it supports the local economy; because it keeps local agricultural land in production, ensuring that future generations will still be surrounded by lots of open fields, grazing lands, and wildlife habitat; because local food increases community food security by retaining the experts that know how to produce food; and because local food has a story—knowing where your food comes from means that its source is not anonymous, but accountable. Lastly, by eating local, you are integrating ecology, community, and gastronomy . . . you are doing well by eating well."

I did well. The food was delicious. I went back for a second burger the next day.

But it's good economic sense too. According to a recent study, while livestock accounted for 93 percent of all agricultural sales in Coconino County, which encompasses Flagstaff, only 0.5 percent of ranch products were sold directly to local consumers. Meanwhile, eaters purchased $37 million of meat, poultry, fish, and eggs from the commodity food system. When the study expanded its analysis to include Navajo, Coconino, and Yavapai counties, it found that only $343,000 of food products were sold directly to consumers versus $635 million of food annually bought from outside sources. *That means roughly $700 million of potential wealth could be captured by local ranches and farmers.*

While a burger joint may not be enough by itself to keep these family ranches in business, it does represent an effort on the part of producers, eaters, and conservationists to try something new under the sun: working together economically.

This is where the hope comes in.

Author and eater Gary Paul Nabhan puts it this way: "You walk away from Diablo Burger with a lingering sense that your decision to eat there has been good for you, for the land, and for the local rural community. What more could you want?"

TWENTY-THREE

REDEFINING LOCAL

(2010)

OKLAHOMA FOOD COOPERATIVE

OKLAHOMA

WHAT DOES "LOCAL" mean exactly when you live on a remote farm or ranch?

It's an important question because no matter where you go today, it seems, "local" is on everybody's lips—and for good reason. Its many advantages address some of the most pressing problems of our time: it gives us access to fresh, healthy food in an economy dominated by industrial agriculture; it reduces our carbon footprint and lessens our dependence on fossil fuels (both of which help fight global warming); it keeps money circulating in the local economy, where its multiplier effect can be significant; it builds a sense of community among all participants; and it pokes globalization in the eye.

But when we talk about "local," we almost always do so from the perspective of the urban dweller, i.e., those products grown or made closest to the customer. Farmers' markets are a good example. "Local" in their case means a radius around a point located in a city or suburb. This means it is self-selecting—it is limited to those farms and ranches that consider themselves to be "local enough" to afford the drive into town every weekend. In other words, from the perspective of a city resident, any farmer selling produce in-person is a "local."

What about all the producers who are not able to make it to a farmers' market but would like to?

If you live on a remote farm or ranch, especially out West where the distances to markets can be staggering, "local" looks very different. You might be able to sell your products in the nearest small town, but this market is likely to be limited in the long run, especially as competition with neighbors, and diesel prices, rise. If a bigger market exists two hours away instead, does that constitute "local?" It's a significant challenge for many rural residents. Without a Santa Fe or Denver or Portland nearby, how can an organic farmer or grass-fed beef rancher participate in the burgeoning local food-and-crafts movement and reap its benefits, especially its profits, if he or she lives way out in the back forty?

Fortunately, the Oklahoma Food Cooperative has come up with an ingenious solution. I think everyone should take a look at what they've accomplished, as I did when I recently drove to western Oklahoma for a tour organized by a few of the producers in the cooperative. What they've come up with is innovative, effective, and (so far) successful. When the cooperative began in 2003, it took thirty-six orders from customers for $3,200 in sales. By 2006, it had nine hundred members, both producers and consumers. Today, it has over two thousand members and does $500,000 in annual sales.

The key to the cooperative's success was a radical idea: they redefined "local" to include the entire state—with significant help from the Internet. Here's what I knew about the cooperative's model before venturing on my field trip:

- All products provided by the cooperative are produced within the state of Oklahoma.
- Beginning on the first day of every month, members can go on the cooperative's Web site and purchase any food or craft product listed.
- Then on the second Thursday, this electronic ordering "window" closes. The orders are then sent to the participating farms and ranches so they can be filled.
- On the third Thursday of the month, designated drivers (usually producers) visit all the participating farms and ranches to pick up the orders.

- All drivers then converge at a warehouse in Oklahoma City, where the products are separated into piles and then rebundled according to the customers' orders.
- The drivers drive back home, dropping off the individual orders at designated locations, where the customers pick them up.

Here are more details:

- The one-time membership fee is $52—the same for producers and customers.
- Each farm and ranch creates its own page on the cooperative's Web site, each sets its own price for its products, each designs its own label and controls the advertising, and each is in charge of its monthly inventory.
- Customers can buy as much or as little as they want each month, the purchase is made through the cooperative, and customers can earn credits toward a purchase by volunteering for the organization at the Oklahoma City Warehouse.
- The cooperative pays every farmer and rancher ninety cents of every dollar spent by the customer; the other ten cents supports the cooperative, which also adds a 10 percent markup on all products (combined, this twenty cents covers the operating expenses of the organization).

Let me repeat that second-to-last point: all farmers and ranchers get ninety cents of every dollar spent on their products. In the industrial agricultural model that dominates food production today, producers typically get nineteen cents of every food dollar. The rest goes to middlemen, including packers, truckers, grocery stores, and other corporate interests. This is one of the reasons why farmers and ranchers have struggled with profitability over the decades. Not only are they required to be "cost takers" from a corporate system that dictates prices for their products (such as feedlot beef), but the few alternatives available to them to increase their cut of every food dollar (such as farmers' markets) have their own challenges.

Of course, it is a little more complicated than this, but the bottom line is that most cooperative producers come out ahead because they are now "price givers" instead of "price takers." They can set their own prices and control, to a certain degree, their costs. This is something relatively new under the sun and is one of the reasons I made the long drive from Santa Fe to check it out.

Another reason was the impressive list of products available each month to members. There are nearly two thousand items on the cooperative's Web site, all made in Oklahoma, and many organic, natural, or grass-fed. A sampling of items include bakery goods, beverages, candy, canned foods, condiments, dairy products and eggs, entrees, fruits, gift boxes, grains, flours and pastas, herbs, jam, and jellies, meats, natural sweeteners, nuts, poultry, prepared foods, side dishes, and vegetables. Also: apparel, art, baby products, bath and beauty supplies, books, classes, fiber arts, fishing supplies, health items, jewelry, laundry care, garden supplies, live plants, and seeds.

The cooperative's model differs in important ways from traditional methods of obtaining local products. For example, members can order what they want, when they want it, and what they can afford, which means they are not locked into the weekly produce list of, say, a Community Supported Agriculture (CSA) farm. No more kale and bok choy this week, thanks! For producers, participation in the cooperative means making only one trip a month into town (and only then if they are a designated driver) instead of the weekly trips required by the farmers' market model. Not only is this easier on the farmer, it's easier on the planet too.

One downside to the cooperative's model, however, is less face-to-face interaction between producers and customers. In both the CSA and farmers' market models, the meet-and-greet relationship between grower and eater is an important part of doing business. By contrast, by working through the Internet, as the Oklahoma Food Cooperative does, growers and eaters don't get much face time (a big problem with the Internet, in my opinion).

But for remote farmers and ranchers, this downside is offset by a big upside: they get to participate in a "local" food economy. By offering products for sale via the Internet at a one-stop shop provided by the cooperative, and then driving to a central hub to distribute the goods, "local" is extended to the state line. Suddenly, "remote" doesn't seem so remote anymore.

It's not as crazy as it sounds. In fact, it's part of a trend. According to the USDA's recent Census of Agriculture, the value of direct farm sales increased 167 percent between 2002 and 2007, which also listed 3,194 farmers as offering direct sales to consumers.

This is great news, and that's another reason I drove to western Oklahoma: to see hope in action. And I found what I was looking for on the very first stop of the tour, at a small farm called Cattle Tracks a mile or so north of Fairview, a certified organic wheat farm and grass-fed beef operation, owned by John and Kris Gosney. Their story was typical of the 125 producer members of the cooperative. Not long ago, John was a conventional wheat farmer, soaking his fields with pesticides, harvesting the wheat with a ton of fossil fuel, and watching his spirit decline along with the land's health. He became depressed, he told the tour group, often finding himself sitting on a bale of hay wondering where his life was heading.

John said that he never gave organic farming a thought until a neighbor asked him to take over his farm, as he was about to retire and didn't want to let his hard work developing an organic wheat operation come to naught. John was immediately struck by the profitability of his neighbor's farm and decided to certify his own farm as organic as well. He saw a drop in yield initially, but he also saw a drop in expenses, especially since he stopped using conventional fertilizers and pesticides. Eventually, as the yield came up, so did his profits.

However, the main benefit of the switch, he said, was noneconomic: he began to have fun again. Going organic cured him of his depression, he explained. He liked the challenge of organic as well as the hard work it requires. A recent musk thistle invasion, for example, necessitates that he spend at least an hour a day with a shovel eradicating the plants. Under the conventional model, of course, he would have sprayed herbicide on the baleful weed.

Today, John and Kris grow cattle to eight hundred pounds on their wheat fields and finish them on native grass (an all-wheat diet influences the taste of the meat, he said). In addition to selling his products through the cooperative, an organic restaurant in Oklahoma City called Sage takes their beef. He proudly pointed to a recent analysis by Oklahoma State University of the CLA (conjugated linoleic acid—a cancer-fighter) content of Cattle Tracks beef. According to the analysis, it was "in the highest range of CLA content reported in the literature for beef."

He also spoke at length about his latest project: brewing microbes in large vats of compost tea. Repeated applications of herbicides and pesticides on his farmland over the decades had effectively destroyed the microbiotic life of the soil. To remedy this, he brewed microbes in a big container in his barn and sprayed them on the land—restoring the natural fertility of the soil organically. As he talked, it was evident how pleased he was with his work.

After the formal question-and-answer period was done, I stepped up and brought up the topic of his depression. "What exactly," I asked him, "makes you happy about organic farming?" He paused and turned inward for a moment. "I feel like I'm finally doing God's work," he said quietly. It was a sentiment, I learned, shared by many on the tour.

To me, John's story is a great example of old-fashioned American know-how in action—applied in this case to the cause of organic farming instead of industrial agriculture. This practical, can-do attitude of farmers, much vaunted over the decades by politicians and others as quintessentially American, was much in evidence on the tour. Unfortunately, when we talk about American know-how today, it is almost always in the context of high technology. Rarely is it discussed in the context of low technology—such as local food systems or organic farming. This is a shame because I believe a great deal of innovative American know-how is alive and well on the back forty—and we should give it a closer look.

I traveled to the next stop on the tour with Kim Barker, a rancher and one of the founders and organizers of the cooperative. He told me that the cooperative, while effective, isn't a cure-all for remote farms and ranches. In fact, some cooperative members still drive long distances to farmers' markets on weekends to sell their products. It's all part of what needs to be done, he said, to make a living as a direct marketer of local food.

Kim is careful to point out that the Oklahoma Food Cooperative is a producer *and* consumer cooperative—not just a collection of farmers and ranchers. In fact, the initial idea for the cooperative came from Robert Waldrop, a "foodie" in Oklahoma City who had a vision for a virtual marketplace that was also locally based. Today, he still serves as president of the enterprise.

"Among our producer and customer members, we find a diversity of lifestyles, beliefs, cultures, and religions," writes Waldrop on the cooperative's Web site. "Even so, we find common ground based on our mutual need for a marketplace where we can find good, healthy nutritious local foods. We are focused on meeting local needs with local resources."

They have succeeded so far because their eyes are firmly fixed on a vision of community, social justice, environmental sustainability, and economically viable local food systems. That vision continues to sustain the cooperative today. And as I mentioned earlier, I'm certain this model can be replicated in any region where there is a need to redefine "local" to include remote farmers and ranchers. Thanks to the Oklahoma Food Cooperative, this vision has become a reality.

TWENTY-FOUR

THE CARBON RANCH

(2011)

> "Carbon is the basic building block for life. It is only a pollutant when in excess in the atmosphere or dissolved in water. Over millennia, a highly effective carbon cycle has evolved to capture, store, transfer, release, and recapture biochemical energy in the form of carbon compounds. The health of the soil, and therefore the vitality of plants, animals, and people, depends on the effective functioning of this cycle."
>
> —DR. CHRISTINE JONES, SOIL SCIENTIST

NOVELIST AND HISTORIAN Wallace Stegner once said that every book should try to answer an anguished question. I believe the same is true for ideas, movements, and emergency efforts. In the case of climate change, an anguished question is this: what can we do right now to help reduce atmospheric carbon dioxide (CO_2) from its current (and future) dangerously high levels?

In an editorial published in July of 2009, Dr. James Hansen of NASA proposed an answer: "cut off the largest source of emissions—coal—and allow CO_2 to drop back down . . . through agricultural and forestry practices that increase carbon storage in trees and soil." I consider these words to be a sort of 'Operating Instructions' for the twenty-first century. Personally, I'm not sure how we accomplish the coal side of the equation, which requires governmental action, but I have an idea about how to increase carbon storage in soils.

I call it a *carbon ranch*.

The purpose of a carbon ranch is to mitigate climate change by sequestering CO_2 in plants and soils, reducing greenhouse gas emissions, and producing co-benefits that build ecological and economic resilience in local landscapes. "Sequester" means to withdraw for safekeeping, to place in seclusion, into custody, or to hold in solution—all of which are good definitions for the process of sequestering CO_2 in plants and soils via photosynthesis and sound stewardship.

The process by which atmospheric CO_2 gets converted into soil carbon is neither new nor mysterious. It has been going on for millions and millions of years, and all it requires is sunlight, green plants, water, nutrients, and soil microbes. According to Dr. Christine Jones, a pioneering Australian soil scientist, there are four basic steps to the CO_2/soil carbon process:

- photosynthesis
- resynthesis
- exudation
- humification

Photosynthesis: This is the process by which energy in sunlight is transformed into biochemical energy, in the form of a simple sugar called glucose, via green plants—which use CO_2 from the air and water from the soil, releasing oxygen as a byproduct.

Resynthesis: Through a complex sequence of chemical reactions, glucose is resynthesized into a wide variety of carbon compounds, including carbohydrates (such as cellulose and starch), proteins, organic acids, waxes, and oils (including hydrocarbons)—all of which serve as fuel for life on Earth.

Exudation: Around 30-40 percent of the carbon created by photosynthesis can be exuded directly into soil to nurture the microbes that grow plants and build healthy soil. This process is essential to the creation of topsoil from the lifeless mineral soil produced by the weathering of rocks over time. The amount of increase in organic carbon is governed by the volume of plant roots per unit of soil and their rate of growth. More active green leaves mean more roots, which mean more carbon exuded.

Humification: This is the creation of humus—a chemically stable type of organic matter composed of large, complex molecules made up of carbon, nitrogen, and minerals. Visually, humus is the dark, rich layer of topsoil that people associate with rich gardens, productive farmland, stable wetlands, and healthy rangelands. Land management practices that promote the ecological health of the soil are key to the creation and maintenance of humus. Once carbon is sequestered as humus, it has a high resistance to decomposition and therefore can remain intact and stable for hundreds or thousands of years.

Additionally, high humus content in soil improves water infiltration and storage, due to its spongelike quality and high water-retaining capacity. Recent research demonstrates that one part humus can retain as much as four parts water. This has positive consequences for the recharge of aquifers and base flows to rivers and streams, especially important in times of drought.

In sum, the natural process of converting sunlight into humus is an organic way to pull CO_2 out of the atmosphere and sequester it in soil for long periods of time. If the land is bare, degraded, or unstable due to erosion, and if it can be restored to a healthy condition, with properly functioning carbon, water, mineral, and nutrient cycles, covered with green plants with deep roots, then the quantity of CO_2 that can be sequestered is potentially high. Conversely, when healthy, stable land becomes degraded or loses green plants, the carbon cycle can become disrupted and will release stored CO_2 back into the atmosphere.

In other words, healthy soil = healthy carbon cycle = storage of atmospheric CO_2.

Any land management activity that encourages this equation, especially if it results in the additional storage of CO_2, can help fight climate change. Or as Dr. Christine Jones puts it, "Any . . . practice that improves soil structure is building soil carbon."

What would those practices be? There are at least six strategies to increase or maintain soil health and thus its carbon content. Three sequestration strategies include:

1. *Planned grazing systems.* The carbon content of soil can be increased by the establishment of green plants on previously bare ground, deepening the roots of existing healthy plants, and the general improvement of nutrient, mineral, and water cycles in a given area. Planned grazing is key to all three. By controlling the timing,

intensity, and frequency of animal impact on the land, a "carbon rancher" can improve plant density, diversity, and vigor. Specific actions include the soil cap–breaking action of herbivore hooves, which promotes seed-to-soil contact and water infiltration; the "herd" effect of concentrated animals, which can provide a positive form of perturbation to a landscape by turning dead plant matter back into the soil; the stimulative effect of grazing on plants, followed by a long interval of rest (often a year), which causes roots to expand while removing old forage; targeted grazing of noxious and invasive plants, which promotes native species diversity; and the targeted application of animal waste, which provides important nutrients to plants and soil microbes.

2. *Active restoration of riparian, riverine, and wetland areas.* Many arroyos, creeks, rivers, and wetlands in the United States exist in a degraded condition, the result of historical overuse by humans, livestock, and industry. The consequence has been widespread soil erosion, loss of riparian vegetation, the disruption of hydrological cycles, the decline of water storage capacity in stream banks, and the loss of wetlands. The restoration of these areas to health, especially efforts that contribute to soil retention and formation, such as the reestablishment of humus-rich wetlands, will result in additional storage of atmospheric CO_2 in soils. There are many cobenefits of restoring riparian areas and wetlands to health as well, including improved habitat for wildlife, increased forage for herbivores, improved water quality and quantity for downstream users, and a reduction in erosion and sediment transport.

3. *Removal of woody vegetation.* Many meadows, valleys, and rangelands have witnessed a dramatic invasion of woody species, such as pinon and juniper trees where I live, over the past century, mostly as a consequence of the suppression of natural fire and overgrazing by livestock (which removes the grass needed to carry a fire). The elimination of over-abundant trees by agencies and landowners has been an increasing focus of restoration work recently. One goal of this work is to encourage grass species to grow in place of trees, thus improving the carbon-storing capacity of the soil. The removal of trees also has an important cobenefit: they are a potential source of local biomass energy production, which can help reduce a ranch's carbon footprint.

Three maintenance strategies that help keep stored CO_2 in soils include:

1. *The conservation of open space.* The loss of forest, range, or agricultural land to subdivision or other types of development can dramatically reduce or eliminate the land's ability to pull CO_2 out of the atmosphere via green plants. Fortunately, there are multiple strategies that conserve open space, including public parks, private purchase, conservation easements, tax incentives, zoning, and economic diversification that helps to keep a farm or ranch in operation. Perhaps most importantly, the protection of the planet's forests and peatlands from destruction is crucial to an overall climate-change-mitigation effort. Not only are forests and peatlands important sinks for CO_2; their destruction releases large amounts of stored carbon back into the atmosphere.

2. *The implementation of no-till farming practices.* Plowing exposes stored soil carbon to the elements, including the erosive power of wind and rain, which can quickly cause it dissipate back into the atmosphere as CO_2. No-till farming practices, especially organic ones (no pesticides or herbicides), not only protect soil carbon and reduce erosion, but they often also improve soil structure by promoting the creation of humus. Additionally, farming practices that leave plants in the ground year-round both protect stored soil carbon and promote increased storage via photosynthesis. An important cobenefit of organic no-till practices is the production of healthy food.

3. *Building long-term resilience.* Nature, like society, doesn't stand still for long. Things change constantly, sometimes slowly, sometimes in a rush. Some changes are significant, such as a major forest fire or a prolonged drought, and can result in ecological threshold-crossing events, often with deleterious consequences. Resilience refers to the capacity of land, or people, to "bend" with these changes without "breaking." Managing a forest through thinning and prescribed fire so that it can avoid a destructive, catastrophic fire is an example of building resilience into a system. Managing land for long-term carbon sequestration in soils requires building resilience as well, including the economic resilience of the landowners, managers, and community members.

All of these strategies have been field-tested by practitioners, landowners, agencies, and researchers and demonstrated to be effective in a wide variety of landscapes. The job now is to integrate them holistically into a "climate-friendly" landscape that sequesters increasing amounts of CO_2 each year.

> "Let's be clear . . . We will still have to radically reduce carbon emissions, and do so quickly. We will still have to eliminate the use of fossil fuels and adopt substantially more sustainable agricultural methods. We will still have to deal with the effects of ecosystems damaged by carbon overload."
>
> —***WALL STREET JOURNAL*, 2009**

Reality check: the increased sequestration of CO_2 in soils won't solve climate change by itself. It won't even be close if the emissions of greenhouse gases are not dramatically reduced at the same time. According to experts, this reduction must be on the order of 50-80 percent of current emissions levels within fifty years. Accomplishing this goal will require a massive rearrangement of our energy sector toward low-carbon technologies as well as big changes in the everyday lives of Americans.

A carbon ranch can help in three ways: by measuring and then reducing the amount of greenhouse gas emissions an agricultural operation contributes to the atmosphere; by producing renewable energy "on-ranch," which it can use itself and/or sell to a local or regional power grid; and by participating in local food and restoration activities that lower our economy's dependence on fossil fuels.

A carbon ranch can also help by confronting the controversy over "offsets" and carbon "credits"—the two strategies most frequently touted by governments, businesses, and others for encouraging the creation of a so-called "carbon marketplace." In this marketplace, "credits" created by the sequestration of CO_2 in one place can be "sold" or traded to "offset" a CO_2 polluting entity, such as a coal plant or airline company, someplace else, supposedly to the benefit of all. In reality, these schemes appear to mostly offset our guilty feelings rather than actually affect atmospheric levels of CO_2.

Here are these ideas in more detail:

Reducing the "footprint" of a carbon ranch. This is a two-step process: assess the amount of greenhouse gas emissions that are rising from a particular landscape or operation, follow this assessment with a concerted effort to reduce these emissions. One way to measure this carbon footprint is to conduct a Life-Cycle Assessment (LCA) of an enterprise, which is an inventory of the material and energy inputs and outputs characteristic of each stage of a product's life cycle. This is a well-recognized procedure for tracking the ecological impacts of, say, a television set or a refrigerator, and different types of LCAs exist for different types of products.

For a carbon ranch, there are four important measures of its LCA:

- cumulative energy use
- ecological footprint
- greenhouse gas emissions
- eutrophying emissions

The first three measurements are relatively straightforward, and there are many credible methodologies today to calculate energy use, ecological footprints, and emissions, though most are designed for urban contexts or industrial agriculture.

However, the fourth measurement—eutrophying emissions—has been the source of considerable controversy in recent years. It refers to the amount of methane produced by the digestive system of livestock during its time on the ranch, farm, or feedlot—and in the public's mind, the connotation is negative. That's because the public has conflated a natural biological process—belching cows—with fossil fuel-intensive industrial livestock production activities, including chemical fertilizer production, deforestation for pasture, cultivation of feed crops (corn), and the transportation of feed and animal products. As a result, there is an impression among the public at large that one answer to the climate crisis is to "eat less red meat"—an opinion that I have heard repeatedly at conferences and meetings.

Personally, I think an answer is to eat more meat—from a carbon ranch.

For the purposes of a carbon ranch, the methane emission issue is just one part of the overall "footprint" assessment. The goal of a Life-Cycle Analysis is to measure an operation's energy use and emissions so that it can reduce both over time. Ultimately, the goal is to become carbon-neutral or, ideally, carbon-negative—meaning the amount of CO_2 sequestered is greater than the ranch's carbon footprint.

Producing renewable energy. Anything that a carbon ranch can do to produce energy on-site will help balance its energy "footprint" and could reduce the economy's overall dependence on fossil fuels. This includes wind and solar farms; the production of biodiesel from certain on-site crops for use in ranch vehicles; biomass for cogeneration projects (this is especially attractive if it uses the woody debris being removed from the ranch anyway); micro-hydro, micro-wind, and solar for domestic use; and perhaps other as yet unrealized renewable energy alternatives.

Participating in a local economy. A carbon ranch should carefully consider its role in the "footprint"of the greater economy. Are its products traveling long distances or otherwise burning large amounts of fossil fuels? It is generally accepted that involvement in a local food market, where the distances between producer and eater are short, shrinks the fossil "footprint" of a ranch considerably. There is some contradictory research on this point, however. In my opinion, the technical issues of local versus global food systems in terms of food miles traveled is largely neutralized by the wide variety of cobenefits that local food brings economically and ecologically.

The trouble with offsets. Many observers—myself included—have become increasingly skeptical of the offset concept at regional or national scales. Objections include:

- We need actual *net* reductions of atmospheric CO_2, not just the neutralizing "offset" of a polluter by a sequesterer. And we need these net reductions quickly.
- It is not acceptable to let a big, industrial polluter "off the hook" with an offset.
- It is unrealistic to expect the same system that created the climate problem in the first place—i.e., our current economy and specifically its financial sector—to solve this problem and to do so with the same financial tools.
- At best, offsets may be illusory; at worst, they're fraudulent—thus imperiling the whole purpose of the approach.

While offsets and carbon credits may not be the economic engine of the future, they highlight an important challenge for carbon ranching: profitability. If not offsets, then how can a landowner who desires to mitigate climate change earn a paycheck, without which there will no carbon ranching?

One idea is to include "climate-friendly" practices as an added value to the marketing of ranch products, such as its beef. Another is to create a "carbon market" at the local level. A county government, for example, could help to create a local carbon market to help offset its judicial buildings or schools or prisons. It could possibly do so through its ability to tax, zone, and otherwise regulate at the county level. It would still have to deal with some of the other challenges confronting offsets, but at least it would keep the marketplace local.

Another idea might be to reward landowners financially for meeting sequestration and emissions goals. The federal government routinely subsidizes rural economic development enterprises, such as the ongoing effort to bring high-speed broadband Internet to rural communities. Additionally, the government often provides incentives to businesses for market-based approaches, including corn-based ethanol production, solar power development, and wind technology (and don't forget the federal government's catalyzing role in the birth of the Internet). It would be perfectly logical, therefore, to reward early adopters of carbon ranching with a direct financial payment as a means to stir up new markets.

None of this will be easy. In fact, the obstacles standing in the way of implementing a carbon ranch and sharing its many cobenefits are large and diverse. Is it worth trying anyway? Absolutely. If a carbon ranch could make a difference in the fight against climate change—now developing as the overarching crisis of the twenty-first century—then we must try. The alternative—not trying—means we consign our future to politics, technology, and wishful thinking, none of which have made a difference so far.

Best of all, a carbon ranch doesn't need to be invented. It already exists. We know how to grow grass with animals. We've learned how to fix creeks and heal wetlands. We're getting good at producing local grassfed food. We'll figure out how to reduce our carbon footprint and develop local renewable energy sources profitably. We don't need high technology—we have the miracle of photosynthesis already.

Answers to anguished questions exist, but too often our eyes seem fixed on the stars and our minds dazzled by distant horizons, blinding us to possibilities closer to home. A carbon ranch teaches us that we should be looking down, not up.

At the grass and the roots.

TWENTY-FIVE

THE FIFTH WAVE

(2012)

"All things alike do their work, and then we see them subside. When they have reached their bloom, each returns to its origin . . . This reversion is an eternal law. To know that law is wisdom."

—LAO-TSU (SIXTH CENTURY, BCE)

THE FIRST WAVE

IN THE FALL of 1909, twenty-two-year-old Aldo Leopold rode away from the ranger station in Springerville, Arizona, on his inaugural assignment with the newly created United States Forest Service. For this Midwesterner, an avid hunter freshly graduated from the prestigious Yale School of Forestry, the mountainous wilderness that stretched out before him must have felt both thrilling and portentous. In fact, events over the ensuing weeks, including his role in the killing of two timber wolves—immortalized nearly forty years later in his essay "Thinking Like a Mountain," from *A Sand County Almanac*—would influence Leopold's lifelong conservation philosophy in important ways. The deep thinking would come later, however. In 1909, Leopold's primary goal was to be a good forester, which is why he chose to participate in a radical experiment at the time: the control and conservation of natural resources by the federal government.

Beginning in 1783, the policy of the federal government encouraged the disposal of public lands to private citizens and commercial interests including retired soldiers, homesteaders, railroad conglomerates, mining interests, and anyone else willing to fulfill America's much-trumpeted manifest destiny. However, this policy began to change in 1872, when President Ulysses Grant signed a bill creating the world's first national park—Yellowstone—launching the U.S. government down a new path: retention and protection of some federal land on behalf of all Americans. In 1891, four years after Leopold's birth, this trend accelerated when Congress created the national forest reserve system, which protected large swaths of valuable timberland from development. These reserves were renamed national forests and were dramatically increased in size in 1907 by President Theodore Roosevelt, who burned the midnight oil with Gifford Pinchot, his visionary secretary of agriculture. Three years earlier, Roosevelt had created the first national wildlife refuge—Pelican Island—in southern Louisiana.

These parks, forests, refuges, and monuments (the latter created by the Antiquities Act of 1906) were part of an audacious conservation philosophy that emphasized state and federal control and scientific management of natural resources. For Pinchot and other leaders in the budding conservation movement, the need for a new approach could be summed up in one word: scarcity. Take timber, for instance. Appalled by the razing of the great white pine forests of the upper Midwest by private industry after the Civil War, Congress created the forest reserve system and gave it the mission of conserving valuable timberlands for future national needs. It was a mission vigorously supported by Pinchot, who believed that a nation's natural resources should serve the greatest good for the greatest number of citizens. This new conservation philosophy was captured in the U.S. Forest Service's first field manual: "Forest Reserves are for the purpose of preserving a perpetual supply of timber for home industries, preventing destruction of the forest cover, which regulates the flow of streams, and protecting local industries from unfair competition in the use of forest and range. They are patrolled and protected, at Government expense, for the benefit of the Community and home builder."

Reversing resource scarcity and arresting the associated land degradation would now be the job of government.

Meanwhile, scarcity of a different sort motivated John Muir, an itinerant mountain lover and amateur geologist from Scotland. Worried about the loss of wildness and beauty to development, Muir campaigned vigorously for the creation of national parks and monuments, adding his voice to what quickly became a chorus of support for the protection of wilderness, wildlife, and natural wonders for nonutilitarian purposes. It worked. The national park system expanded from two dozen units in 1916—the year Congress created the National Park Service—to over four hundred only eight decades later. The federal role in the West continued to expand after World War II, when the vast public rangelands were organized into the Bureau of Land Management (BLM). In 1964, Congress added an additional layer of protection with the passage of the Wilderness Act, which ensured that roadless areas on public lands would remain "untrammeled" for generations to come.

It was all part of the first wave of conservation, which I'll call *federalism.*

These were heady days for professionals such as Leopold, but also exciting times for day-trippers and vacationers across the nation, newly liberated by rising affluence and declining prices of automobiles. Recreation quickly took its place alongside resource protection as part of the mission of federal land agencies. Starting in the 1920s, America embraced its parks and forests with fervor as citizens hit the roads in rising numbers. In the process, a benevolent and ever-helpful "Ranger Rick" became synonymous with the U.S. government in the public's eyes.

Meanwhile, the nation's embrace of the great outdoors had an important collateral effect: federalism as a conservation philosophy began to extend beyond land ownership and management to the belief that governmental regulation of the environment was needed in order to protect citizens from harm. Thanks to pressure from activists, more and more regulatory work was assigned to the federal government over the decades, culminating in the creation of the Environmental Protection Agency (EPA) in 1969 and a raft of historic environmental legislation in the early 1970s.

Federalism, it seemed, was destined to keep rolling ashore.

Today, however, it is clear that this first wave of conservation has faded. In retrospect, its apogee as an effective conservation strategy in the West was reached in the early 1950s, just prior to the eruption of major controversies involving the government's dam-building program on the Colorado River and its over-harvesting of timber on our national forests—controversies that began to sour the public on some of our federal agencies. This souring mood grew during the 1960s and 1970s as activists fought the government over hard-rock mining, cattle grazing, and endangered species protection on public lands, causing many urban residents to shift their view of federal agencies from the good guys to the bad guys. It was a shift shared by many rural residents, who began to view the government as captive of urban interests, environmental activists especially. As a result, federal employees began to find themselves in the crossfire of an increasingly rancorous struggle between activists and rural residents across the West. It added up to one inescapable conclusion: federalism as an effective conservation strategy was fading away.

That's not to say the *idea* of public land staled—the democratic ideal represented by public ownership of Western lands is still strong. What has changed is the government's ability *to do* conservation effectively. It has faded in recent years for a variety of reasons, including shrinking budgets, reduced personnel, increased public demands, a bevy of conflicting laws and regulations, and the rising hostility of political interests. But the conservative and conformist nature of bureaucracies had a role too. Over time, a resistance to innovation grew among the agencies, as did a certain degree of arrogance. Toss in a lack of synchronicity with the times, as public opinions changed and new ideas came along, and by the 1970s, the result was increased ineffectualness. Not that federalism didn't try to evolve with the times. Over the years, it embraced a variety of new conservation concepts, including wilderness protection, sustained yield, adaptive management, endangered species protection, an ecosystem approach, and so on. But none of them altered the fact that what had once been federalism's chief asset—its role as a buffer between nature and its exploiters—had by the 1970s become its chief liability: it now stood between the land and innovation.

I experienced this firsthand with Quivira's work with federal land agencies, including our promotion of progressive livestock management, our direction of riparian restoration projects, and our operation of the only public lands grassbank in the West (where Quivira became a Forest Service livestock permittee). I'll cite three examples. First, it became clear that the default position of agencies on anything out of the box was "no"—no to this idea, no to that activity; no, you can't do this; no, you can't do that. Getting to "yes" wasn't impossible with the agencies, but their regulatory mandates, musical-chair personnel changes, and ever-rising workloads make getting to "yes" a time-consuming, expensive, and very frustrating process for potential partners. It is much simpler for the federal agencies to say "no."

Second, there were few positive internal incentives for agency employees to try anything new. In fact, disincentives abounded, including the perpetual threat of lawsuits by watchdog groups. Innovating within the system is rarely rewarded and sometimes punished. Thinking out of the box might mean getting pushed out of your job. There is much less stress for employees if they act by the book—which often made partners feel like they were talking to a stone wall.

Third, there is a culture of command and control within the federal agencies, the Forest Service especially, that discouraged partnerships and innovation. Agencies often have the last word on a project, and they know it. This means that when they enter into a collaborative effort, the partnership is unequal. The agencies have the ability to shut things down, and all it takes is one person in a position of power. Throw in the inevitable change of leadership among line officers every three to four years, and the risk of "no" rises substantially. For example, of the approximately twenty Forest Service employees involved in the creation of the grassbank in 1998, nineteen had moved to new jobs within five years, essentially orphaning the project from the government's perspective.

It all adds up to an ineffective Status Quo on public lands today. The trouble is that in the twenty-first century, the Status Quo isn't really an option anymore. Managing land for climate change, for instance, will require rapid, flexible, and innovative responses—a tall order for federal agencies. To their credit, agencies sense this and are trying to find ways to respond, but reform, innovation, and breaking gridlock look largely out of their reach now. Perhaps federalism will reinvent itself, gather strength, and rise again as a new wave of conservation. I hope so. There is still a big need for federal oversight and expertise, and the idea of public land ownership is an important one in a democracy.

THE SECOND WAVE

THE NEXT WAVE of conservation in the American West is what we today call *environmentalism*. The early stirrings can be traced back to the mid-nineteenth century as the destructive effects of the Industrial

Revolution began visibly to impact the natural world, especially wildlife populations. Early prophets included Henry David Thoreau, George Perkins Marsh, and John Muir. A vocal advocate for federalism, Muir also played a key role in the development of the second wave when he founded the Sierra Club in San Francisco in 1892. Initially a hiking and camping association for outdoor enthusiasts, the Sierra Club quickly drew activists into its fold, no doubt inspired by Muir's relentless campaign to protect Yosemite National Park from a proposed dam in Hetch Hetchy Valley (a dam that Gifford Pinchot enthusiastically supported). Although Muir lost the fight, his defeat propelled the Club and other budding conservation organizations to become vigilant in defense of the nation's parks, forests, and refuges—and to keep a watchful eye on the federal agencies entrusted to protect them.

As the nation's love affair with the great outdoors took off, conservation groups swelled with new members and advocates, beginning a period of vigorous activity, including a highly public fight in 1955 to stop another dam project, this one located in Echo Park, deep inside Utah's Dinosaur National Monument. Led by the Sierra Club's president, David Brower, an avid mountain climber, the conservation community set itself squarely against Congress and the federal government in a high-stakes showdown. It won. The dam was never built. Riding the momentum of this victory, the second wave swelled in 1963 with the publication of Rachel Carson's *Silent Spring*, which propelled activists into the arena of human health and industrial pollution, transforming the conservation movement into what today is simply called *environmentalism*.

There are two principle reasons why this movement grew large and effective: (1) it built on the strengths of federalism while confronting its weaknesses, and (2) it synchronized itself with the rapidly changing times, including changing demographics, embracing new ideas and values, and putting them to work effectively.

Although the early phase of the second wave was consonant with the goals of federalism, especially the push to create new parks and monuments, as early as the 1930s, it started to have doubts about governmental effectiveness. Led by Aldo Leopold, who had left Forest Service employment in 1924, conservationists began to question the ability of agencies in the wake of the Dust Bowl to implement what Leopold later dubbed a "land ethic." Some government programs worked, but many did not, especially after the positive incentives they employed (direct payments to landowners, technical assistance, etc.) ended. That left many agencies holding the "stick" approach to conservation, rather than the "carrot." However, Leopold came to believe that both approaches were ineffective in the long run because a land ethic needed to come from the heart, not a bureau. He urged the conservation movement to lift its sights to change America's ethics, not just its policies.

Activists responded vigorously to Leopold's call, and environmentalism swelled, especially as America's economy rocketed into the stratosphere after World War II. They began by pushing federal agencies to adopt higher environmental standards. Activists raised alarms, for example, when the Forest Service embarked on a vast timber-cutting program in the 1950s that included widespread clear-cuts. They also criticized the BLM for its poor oversight of livestock grazing and hard-rock mining on public lands and they maintained their struggle with the Bureau of Reclamation, winning a widely publicized fight to stop two dams in the bottom of the Grand Canyon. They also criticized the U.S. Fish and Wildlife Service for its inadequate oversight of endangered species, and they even turned up the heat on the National Park Service, which they thought was dragging its feet on wilderness designation.

In this work, the second wave both shaped public opinion concerning environmental protection as well as followed its lead. In the mid-1960s, a series of natural disasters and slow-boil crises caught the public's attention, including smog in big cities, toxic-waste dumps, oil spills, rivers catching on fire, urban sprawl, and a growing concern about nuclear power. The consequence of this rising concern was the passage of a raft of federal legislation in the early 1970s aimed at ensuring clean air, clean water, endangered species protection, wild and scenic river designation, and an open planning process for the management of public land.

Environmentalism also tapped into changes on the economic front out West, as recreation and tourism became significant engines of prosperity—a development that would eventually be called the "New West." It was a booming amenity-based economy that emphasized recreation (hiking, fishing, biking) over traditional forms of work (mining, logging, farming, cattle ranching). However, the denigration of work in favor of play, especially on public lands, led to numerous clashes with rural residents, many of whom staunchly opposed this new economy. Feelings on both sides hardened during the 1980s, causing environmentalists to dig in and redouble their efforts, which proved successful on many fronts.

In reality, it was a sign of the wave's inevitable fading.

Today, despite environmentalism's continued hard work, high profile, and large memberships, it is clear that the movement is no longer an effective conservation strategy in the West. Two important metrics support this observation: (1) the continued steady decline of animal and plant species populations and their habitats around the planet, and (2) a steady loss of interest in nature and outdoor activities among Americans, especially the younger generation, a trend with alarming ramifications for both nature and people—a condition that author Richard Louv calls "nature deficit disorder." Environmentalism didn't cause these two developments, of course, but it has become increasingly ineffective at reversing, or even curbing, them. There are three primary reasons why.

The first is author and farmer Wendell Berry's long-standing criticism that environmentalism never developed an economic program to go along with its preservation and health programs. It had no economic retort, in other words, for industrialism. It never truly confronted our economy, the source of most environmental ills, and without an effective alternative, the average American had no choice but to participate in a destructive model of economic growth. I saw this played out during my time in the Sierra Club, where I learned that most activists considered environmental problems to have environmental solutions, ignoring their economic sources. This meant we spent too much time and energy on symptoms instead of causes. Aldo Leopold flagged this problem decades earlier when he cautioned us against trying to "fix the pump without fixing the well." We didn't heed his advice, and for fifty years, we focused our attention on the pump while the well began to run dry.

Many environmentalists might argue, in contrast, that they did have an economic agenda: tourism and recreation. This is true—and for a while, the benefits of both looked generous. But over time, recreation and its associated side effects—congestion, exurban sprawl, transitory populations—began to take on darker hues and may have even made the situation worse. As the twenty-first century progresses, with its concerns about climate change, carbon footprints, oil depletion, food miles, and sustainability in general, an economy based on tourism looks increasingly shaky.

Second, environmentalism is ebbing because it left the land behind. The movement lost the feeling of "the soil between our toes," as Leopold put it, meaning it lost an intimate understanding of how land actually works. As a result, it lost what Leopold described as the role of individual responsibility for the health of the land. "Health is the capacity of the land for self-renewal," he wrote, and "conservation is our effort to understand and preserve this capacity." But by losing the feel of soil between our toes, the movement missed the opportunity to understand, and thus preserve, land health—the foundation on which all health depends.

For example, I learned early in my work with Quivira that while activists and others could recognize poor land use, such as overgrazing, and rightly worked to correct it, they lost an understanding of *good* land use, particularly those for-profit activities such as logging and ranching that could be conducted sustainably. Instead, as the movement drifted away from land, it began to equate non-use with the highest and best use of land, especially on the public domain. The exception was recreation, of course, though it has become increasingly clear that as far as twenty-first-century challenges go, play can't handle the weight.

Third, the environmental movement never really walked the walk of a land ethic. While trumpeting Leopold's famous call to enlarge our ethical sphere to include plants and animals, environmentalists ignored his insistence that people and their economic activities be included too. "There is only one soil, one flora, one fauna, and one people, and hence only one conservation problem," Leopold wrote in *A Sand County Almanac*. "Economic and esthetic land uses can and must be integrated, usually on the same acre." Or this from his essay, "The Ecological Conscience": "A thing is right only when it tends to preserve the integrity, stability, and beauty of the community, and the community includes the soil, waters, fauna, and flora, as well as people."

A land ethic encompassed it all. But environmentalists didn't heed Leopold's advice. Instead, many engaged in a form of environmental isolationism. Work was segregated from nature, and nature was largely confined to parks, wildernesses, refuges, and other types of protected areas. Not only was there no attempt to integrate people into nature economically under this preservationist paradigm, but an energetic effort was made by some activists to curtail certain land uses, such as ranching, whether they maintained the integrity, stability, and beauty of the community or not. The land, in their minds, had to be saved apart from the people, and their pitch to the public emphasized dehumanized landscapes—pretty pictures of wild country and charismatic wildlife. In general, while activists were quick to invoke Leopold in their campaigns to save this or that, they ignored his holistic view that "bread and beauty grow best together."

In its time, environmentalism accomplished an astonishing amount, and the world has benefitted immensely from its diligent efforts. As with federalism, however, it reached its "bloom" and began to fade away.

THE THIRD WAVE

THE NEXT WAVE of conservation, which stirred after World War II, had two principal components: an emphasis on science and a focus on private land. This was no accident—these components represented important shortcomings of the previous two waves. Federalism, by definition, focused on public lands, which meant that one-half of the American West—its privately owned land—had been largely neglected by the conservation movement. This became a pressing concern after the war as the suburban and exurban development of private land sped up considerably. Meanwhile, the rise of ecology and other environmental disciplines meant that data and scientific study could now complement, and sometimes supplant, the emotional and romantic nature of environmentalism. An illustrative example is the rise and growth of the Nature Conservancy, a landmark nonprofit organization that is now one of the largest conservation groups in the world.

In 1946, a small group of scientists in New England formed an organization called the Ecologists Union with the goal of saving threatened natural areas on private land, especially biological hot spots that contained important native plant and animal species. The protection of biologically significant parcels of land had traditionally been the job of the federal government, state wildlife agencies, or private hunting and fishing groups. Parks, forests, refuges, wilderness areas, and game preserves were the dominant means by which protection was provided to critical areas in the years leading up to World War II. But a growing number of scientists believed this strategy wasn't sufficient any longer because it largely overlooked privately owned property—land that was rapidly being paved over in the postwar boom.

The Ecologists Union changed its name in 1951 to the Nature Conservancy (TNC) and embarked on a novel strategy: private land acquisition for ecological protection. In 1955, the organization made its first purchase—sixty acres along the New York–Connecticut border. Six years later, it donated its first conservation easement, which restricts development rights on a property in perpetuity, on six acres of salt marsh, again in Connecticut. This new strategy of buying and preserving land caused the organization to grow rapidly. By 1974, TNC was working in all fifty states, often in tandem with state and federal agencies. It wasn't all about acquisition, however. Frequently, TNC acted as the middleman buyer between a willing seller and the federal government. In the process, TNC became adept at real estate deals, developing a business acumen that was as novel for a conservation organization at the time as was its land-protection strategy. TNC also started an ambitious land trust program to accept conservation easements on property it did not own.

Soon, TNC was working internationally, buying land and facilitating major conservation projects. In 2000, it launched the "Last Great Places" campaign, raising over one billion dollars for land acquisition and research. By 2007, TNC was protecting more than 117 million acres of land and five thousand miles of rivers in the U.S. alone.

But it wasn't just about buying land. Employing hundreds of scientists, TNC has based much of its conservation work on research, including a science-based modeling approach to large landscapes that helps the organization determine where to work, what to conserve, and what strategies should be employed. Their work was no longer simply focused on saving the rarest species here and there, as it had been in the 1950s. Now they worked at the ecosystem level across a large landscape so that all species might thrive—a strategy TNC calls "enough of everything." They do this by establishing science-based priorities and then setting out to influence the social, political, and economic forces at work in these biologically important landscapes.

TNC's approach has been replicated by many other third-wave conservation organizations, including Conservation International, the Trust for Public Land, and the World Wildlife Fund. It also helped to ignite a land trust movement around the world. Today, there are over seventeen hundred individual land trusts in America alone, focused on private property of every shape and size, from small community or regional trusts to statewide agricultural organizations.

A great deal of science-based conservation work was also integrated into various nonprofit organizations, public agencies, and private operations. The growing impact of ecology in conservation during the 1940s—thanks in no small part to Aldo Leopold—also led schools and universities to embrace science-based curriculums and implement numerous environmental-study programs across the country. Professional journals in ecology proliferated as a result. At the same time, many public lands–focused environmental organizations incorporated science into their advocacy work, especially those focused on saving large predators, wildlife corridors, and endangered species.

In contrast to environmentalism, however, the third wave eschewed the noisy emotionality and confrontational tactics of the second wave, preferring the quiet diplomacy of research and deal making to accomplish its goals. Although it still adhered to a protection paradigm that it shared with the first two waves, it was guided by data, not poetry, and it sought cooperation, not regulation or litigation, to accomplish its objectives. And as the success of TNC demonstrates, this wave was extraordinarily effective—for a while.

The bloom began to fade in 1990, when TNC purchased the beautiful and biologically rich 322,000-acre Gray Ranch, located in the boot heel of southwestern New Mexico. Sheltering more than seven hundred species of plants, seventy-five mammals, fifty reptiles, and 170 species of breeding birds, the Gray Ranch was considered one of the most significant ecological landscapes in North America, which is why the U.S. Fish and Wildlife Service had coveted the Gray as a wildlife refuge for decades. Indeed, in the 1980s, a similar-sized ranch in southern Arizona, called the Buenos Aires, was purchased by the U.S. Fish and Wildlife Service from the same Mexican millionaire who owned the Gray Ranch. This time, however, the financial terrain was different, and TNC was needed to broker a deal, which it did at a high financial cost to the organization. No matter—TNC had every intention of quickly reselling the Gray Ranch to the federal government and recouping its investment.

Except the transfer never took place.

When local residents heard of the Gray's purchase and pending resale to the federal government, they raised vigorous objections. Going first to their elected representatives and then to the media, their opposition became front-page news across the West, and for a reason: it fit a changing mood in the region. Across the West, pushback against federalism and environmentalism had been gathering steam, often expressed noisily as an exercise of private property-rights. It was more complicated than that, of course, but the bottom line was the same: push had come to shove in the rural West. The Animas-area residents raised three objections to what TNC was trying to accomplish:

(1) the Gray was still a working cattle ranch and thus a tax-paying, cowboy-hiring member of the local economy, and residents wanted it to stay that way; (2) a wildlife refuge would destroy the cultural and historical significance of the Gray, which was part of the historic Diamond A ranch, one of the area's legendary operations; and (3) it was time to stop this pattern of transferring private land to the federal government.

It was this latter point that made the headlines.

Local residents took their complaints directly to TNC officials where, to their surprise, they found a sympathetic reception. That's because TNC was hearing similar complaints in other places around the West. It gave the organization pause—not simply because they didn't like controversy, but because TNC had always considered itself to be a *cooperative* conservation group. Their method was to buy land and easements from willing sellers, to work collaboratively with government agencies, and to create deals that benefitted people and nature while keeping a low profile. But local residents disagreed, saying TNC was *not* being cooperative—not with them, anyway. The complaints stung, causing TNC to ask itself an important question: could it accomplish its scientifically guided conservation goals while maintaining the Gray Ranch as a privately owned working cattle ranch? And perhaps just as importantly: could it find a conservation buyer who would help them recoup their substantial financial stake in the property?

The answer to both questions proved to be "yes."

In 1993, the Nature Conservancy sold the Gray Ranch to Drum Hadley, a local rancher who also happened to be an heir to the Budweiser beer fortune. After the sale, Hadley and members of his family created the Animas Foundation, named for the nearest town, to manage the ranch for conservation as well as community goals. That seemed like a contradiction to many environmentalists, who subsequently objected to TNC's new plan, though to no avail. It all added up to a new approach toward conservation. Success would require that TNC, the Gray Ranch, local residents, and public agencies effectively cooperate together. To that end, a year later, TNC and the Animas Foundation became charter members of the Malpai Borderlands Group, a pioneering collaborative partnership of ranchers, conservationists, and government agencies in the region—setting the stage for the next wave of conservation in the West.

The third wave faded for two reasons mainly: first, the benefits of a protection paradigm, whether science based or not, grew less effective over time as environmental troubles diversified. Climate change, for instance, largely defies the paradigm—what does "protection" mean under rising temperatures, water scarcity, and climatic disorder? Piecemeal protection also exposed the paradigm's limitations as subdivision developments boomed across the West. TNC and other organizations were confronted with a growing dilemma: What benefit is there in buying a large property for protection purposes if the neighboring ranches sell out to a subdivider, thus fragmenting the surrounding land? Also, the top-down approach of the third wave, which shared a command-and-control philosophy with federalism and environmentalism, met increasing resistance from bottom-up groups, limiting its effectiveness. Locals wanted to be heard and involved now. Directives by outsiders, no matter how well-meaning, provoked pushback among the grassroots.

Second, this wave failed to develop a viable economic program to go along with its protection paradigm. While supportive of working landscapes, it struggled to help local residents find paychecks in conservation-friendly enterprises. For example, while TNC could afford to manage its own land without a profit motive, it had great difficulty finding an economic strategy that would keep its neighbors in business (and thus keep "For Sale" signs from appearing). As the subdivision crisis in rural counties heated up in the 1990s, TNC realized that it could not buy all the critical land needed to protect species. There simply wasn't enough money. Nor would conservation easements complete the job. Some sort of conservation economy would be necessary—other than tourism and recreation. To this end, TNC tried a variety of economic strategies, including a "Conservation Beef" pilot project in Montana, but it wasn't enough. Despite TNC's success, it became clear to many that in order to accomplish the landscape-scale effort needed to help species *and* local people, especially if it involved public lands, a new approach would be required, one that featured partnerships and profits.

THE FOURTH WAVE

IN 1991, THE Forest Service extinguished a five-hundred-acre fire burning on private land along a stretch of the remote Geronimo Trail Road, located in the southeastern corner of Arizona. On the surface, it was an unremarkable event—the Forest Service had long reacted to wildfires with the same response: put it out. Period. Except this fire proved to be different. The local ranchers did not want it extinguished, agreeing with scientists that fire had an important role to play in ecosystem health. They asked the federal government to let the fire burn, arguing that it posed no appreciable threat to life or property. The landowner was supportive too; in fact, he had thinned the overgrown brush recently in order to create the right conditions for fire's return. But the Forest Service didn't listen. It put the fire out over all protest. This routine act, however, ignited the community into action. "No more," it said aloud. Consequently, within three years, the nonprofit Malpai Borderlands Group was born. They were determined to do things differently within the nearly one-million-acre borderland they called home. They decided to give collaboration a try.

It was a similar story around the West at the time. When a federal judge shut down logging in old-growth forests throughout the Pacific Northwest in 1991 in response to a lawsuit by environmentalists over the spotted owl, it ignited a storm of protest in rural communities. It also lit two small, but important, bonfires of change. The first was in the Applegate Valley of southwestern Oregon, where a small coalition of activists, loggers, and Forest Service personnel met for potluck suppers and peacemaking. The second was a similar group that met in the only place they considered neutral in the logging-dependent town of Quincy in Northern California—the public library. The goal of both groups was the same: better forest management through collaboration, not confrontation.

In Montana, the Malpai Borderlands Group quickly inspired two groups of ranchers to give collaboration a try, one in the Blackfoot River Valley northeast of Missoula, and the other in the Madison Valley, northwest of Yellowstone National Park. Like Malpai, residents in both valleys grappled with a host of challenges, including the threat of land fragmentation due to subdivisions, curtailment of livelihoods due to endangered species regulations, and changing demographic trends. Instead of fighting the future, however, they chose to link arms with conservationists, scientists, and agency employees with the goal of making progress where it mattered: on the ground. It wasn't easy, especially in the beginning. In many places, trust had to be rebuilt or created; in others, key players wouldn't come to the table. This changed over time, however, as people began to see genuine results. The process was messy, difficult, time-consuming, and frustrating, but it worked.

One name for this new wave is the "radical center"—a term coined by rancher Bill McDonald of the Malpai Borderlands Group. It was radical because it challenged various orthodoxies at work at the time, including the belief of environmentalists that conservation and ranching were part of a zero-sum game—that one could only advance if the other retreated. The "center" referred to the pragmatic middle ground between extremes. It meant partnerships, respect, and trust. But most of all, the center meant *action*—a plan signed, a prescribed fire lit, a workshop held, a hand shook. Words were nice, but working in the radical center really meant *walking the walk.*

I know because I did a lot of the walking myself.

The fourth wave drew strength from the first three waves, while filling in blanks and correcting important deficiencies. It aimed to protect open space and wildlife, valued working landscapes, incorporated public lands, employed ecology and other sciences, and required trust and fairness. But it also strove toward economic realities, often by exploring and promoting the diversification of business enterprises on private lands.

In doing this work, the fourth wave emphasized profits along with protection, arguing persuasively—as Aldo Leopold tried to do years earlier—that good stewardship flowed from ethical and regenerative attitudes toward land, business, and people. Profit could be a force *for* conservation, the fourth wave said, not *against* it, as so many environmental activists had insisted. The proof was in the pudding of these early collaborative efforts: conservation and capitalism (of the local sort) worked effectively side-by-side across the West. The keys were partnerships and dialogue—handshakes and countless meetings. It all led to a rapid expansion of collaboratives of varying stripes in the late 1990s, including the formation of many watershed-based nonprofit organizations. The radical center united, rather than divided.

One area where it worked best was ecological restoration. Ecology had led to a deeper understanding of land sickness—to use Leopold's term—and what to do to restore forests, rangelands, and riparian areas back to health. Ranchers, conservationists, agency personnel, and others began to implement these ideas in pilot projects around the region, including the use of livestock to control noxious weeds, riparian and upland restoration work for water-quality and wildlife-habitat improvement, tackling forest overgrowth through thinning and prescribed fire, and repairing and upgrading low-standard roads in order to restore natural hydrological cycles. Success, however, required cooperation among multiple stakeholders, particularly across private/public and urban/rural divides.

For all its success, however, the fourth wave will too, in time, begin to fade. As the wave evolved from its gridlock-breaking and peacemaking roots into an effort that has brought ecological and economic health to the region and its people, the world evolved too, bringing with it new challenges and opportunities. In short, the times are changing again, especially as we enter into a period of increased climate instability and economic stress.

THE FIFTH WAVE

> "The agrarian population among us is growing, and by no means is it made up merely of some farmers and some country people. It includes urban gardeners, urban consumers who are buying food from local farmers, organizers of local food economies, consumers who have grown doubtful of the healthfulness, the trustworthiness, and the dependability of the corporate food system—people, in other words, who understand what it means to be landless."
>
> —WENDELL BERRY

I TRAVELED UP New York's Hudson Valley to visit a young leader of the emerging agrarian movement by the name of Severine von Tscharner Fleming. I had met Severine a few times before, and I knew her to be an astonishingly energetic and successful advocate for young farmers like herself. For starters, in 2007, she founded the Greenhorns, a nonprofit organization that has become an influential grassroots network dedicated to recruiting and supporting young farmers and ranchers. Severine also cofounded the National Young Farmers Coalition, manages a weekly radio show on Heritage Radio Network, writes a popular blog, speaks at countless conferences, and organizes endlessly via the Web. *And* she's a farmer too.

Severine told me young people are inspired to get into farming for a wide variety of reasons. It starts typically with a journey through apprenticeships and internships as each young farmer discovers which parts of a farming life he or she wishes to pursue, followed by hard work to gain proficiency in, say, carpentry, horse wrangling, or irrigation system maintenance, without going into debt, and usually before starting a family.

Who are these young farmers? According to Severine, most are from cities and suburbs—thus the "greenhorn" moniker—and many come from the social justice or food poverty movements. Another portal is the Food Corps, which is a project of AmeriCorps and places young people in food-oriented jobs, often building school gardens. Many young farmers attended farms when they were kids or went on field trips to local farms through their elementary schools. A few participated in 4-H, though not as many as one might think, she said. The educational backgrounds of young farmers today varies widely, including engineering, public health, computer science, literature, anthropology, and earth science, but the decision to go into farming after examining all the options is the same: to live a life with dignity and purpose and have a positive impact on the community.

"We'll seize opportunities to buy inexpensive battered pastures and compacted soils," she said at a conference, "and then heal those lands using good land stewardship techniques. We'll reclaim territory from commodity crops and try our best not to churn or ruin our own soils while we build up enough capital to stop rototilling. We'll process our own darn chickens and build our own darn websites. We are just as stubborn and innovative as farmers have always been."

According to the USDA Agricultural Census, the number of young people farming in the U.S. is on the rise. Though it is still a minority of the tiny minority of Americans who are farmers, it reinforces the argument that a movement is growing, called by many a *New Agrarianism*.

What does "agrarian" mean exactly? In Latin it means "pertaining to land." My dictionary defines it as relating to fields and their tenure or to farmers and their way of life. Berry broadens this definition, calling it a way of *thought* based on land—a set of practices and attitudes, a loyalty and a passion. It is simultaneously a culture and an economy, he says, both of which are inescapably *local*—local nature and local people combined into "a practical and enduring harmony." The antithesis of agrarianism is industrialism, which Berry says is a way of thought based on capital and technology, not nature. Industrialism is an economy first and foremost, and if it has any culture, it is "an accidental by-product of the ubiquitous effort to sell unnecessary products for more than they are worth."

An agrarian economy, in contrast, rises up from the soils, fields, woods, streams, rangelands, hills, mountains, backyards, and rooftops. It embraces the coexistences and interrelationships that form the heart of resilient local communities and local watersheds. It fits the farming to the farm and the forestry to the forest. For Berry, the agrarian mind is not regional, national, or global, but local. It must know intimately the local plants and animals and local soils; it must know local possibilities and impossibilities. It insists that we should not begin work until we have looked and seen where we are; it knows that nature is the "pattern-maker for the human use of the earth," as he describes it, and that we should honor nature not only as our mother, but as our teacher and judge.

I first ran across the term New Agrarianism in 2003 in a book of essays on the topic collected and edited by Eric Freyfogle, a law professor at the University of Illinois. The term resonated with me because it described exactly what I was seeing on the land. In fact, I could have used Freyfogle's own words from his essay "A Durable Scale" to describe my experience. "Within the conservation movement," he wrote, "the New Agrarianism offers useful guiding images of humans living and working on land in ways that can last. In related reform movements, it can supply ideas to help rebuild communities and foster greater virtue. In all settings, agrarian practices can stimulate hope for more joyful living, healthier families, and more contented, centered lives."

In his essay, Freyfogle produced a list of New Agrarians that was spot on:

- The community-supported agriculture group that links local food buyers and food growers into a partnership, one that sustains farmers economically, promotes ecologically sound farm practices, and gives city dwellers a known source of wholesome food.
- The woodlot owner who develops a sustainable harvesting plan for his timber, aiding the local economy while maintaining a biologically diverse forest.
- The citizen-led, locally based watershed restoration effort that promotes land uses consistent with a river's overall health and beauty.
- The individual family, rural or suburban, that meets its food needs largely through gardens and orchards, on its own land

or on shared neighborhood plots, attempting always to aid wildlife and enhance the soil.

- The farmer who radically reduces a farm's chemical use, cuts back subsurface drainage, diversifies crops and rotations, and carefully tailors farm practices to suit the land.
- The family—urban, suburban, or rural—that embraces new modes of living to reduce its overall consumption, to integrate its work and leisure in harmonious ways, and to add substance to its ties with neighbors.
- The artist who helps residents connect aesthetically to surrounding lands.
- The faith-driven religious group that takes seriously, in practical ways, its duty to nourish and care for its natural inheritance.
- The motivated citizens everywhere who, alone and in concert, work to build stable, sustainable urban neighborhoods; to repair blighted ditches; to stimulate government practices that conserve lands and enhance lives; and in dozens of other ways to translate agrarian values into daily life.

To this list I could add from my recent research:

- The carbon farmer or rancher who explores and shares strategies that sequester CO_2 in soils and plants, reduces greenhouse gas emissions, and produces cobenefits that build ecological and economic resilience in local landscapes.

Freyfogle shares Berry's belief that agrarianism is the proper countervailing force to industrialism and its surfeit of sins, including water pollution, soil loss, resource consumption, and the radical disruption of plant and wildlife populations—the focus of the earlier waves of conservation. Freyfogle goes on to add broader anxieties: the declining sense of community; the separation of work and leisure; the shoddiness of mass-produced goods; the decline of the household economy; the alienation of children from the natural world; the fragmentation of neighborhoods and communities; and a gnawing dissatisfaction with core aspects of our modern culture, particularly the hedonistic, self-centered values and perspectives that control so much of our lives now.

In contrast to these negative attributes of modern life, the new agrarianism is first and foremost about living a life of positive energy and joy, says Freyfogle. Nature is the foundation of this joy, but so are the skills necessary to live a life. At its best, the agrarian life is an integrated whole, with work and leisure mixed together, undertaken under healthful conditions and surrounded by family.

"When all the pieces of the agrarian life come together," Freyfogle wrote, "nutrition and health, beauty, leisure, manners and morals, satisfying labor, economic security, family and neighbors, and a spiritual peacefulness—we have what agrarians define as the good life."

And it is to this good life that the fifth wave aspires.

I credit Aldo Leopold for laying the foundation for this resurgent agrarianism. Over the course of a diverse and influential career, Leopold eloquently advocated a variety of critical conservation concepts, including wilderness protection, sustainable agriculture, wildlife research, ecological restoration, environmental education, land health, erosion control, watershed management, and, famously, a land ethic. Each of these concepts resonates today—perhaps more so than ever, as the challenges of the Age of Consequences grow more complicated and more pressing. But it was Leopold's emphasis on conserving whole systems—soil, water, plants, animals, and people together—that is most crucial today. The health of the entire system, he argued, is dependent on its indivisibility, and the knitting force was a land ethic—the moral obligation we feel to protect soil, water, plants, animals, and people together as one community.

After Leopold's death in 1948, however, the idea of a whole system broke into fragments by a rising tide of industrialization and materialism. Fortunately, today a scattered but concerted effort is underway to knit the whole back together, beginning where it matters most—on the ground. Leopold's call for a land ethic is the root of a New Agrarianism—a diverse suite of ideas, practices, goals, and hopes all based on the persistent truth that genuine health and wealth depends on the land's fertility.

The New Agrarians practice what Aldo Leopold called a unifying force, something "that reaches into all times and places, where men live on land, something that brackets everything from rivers to raindrops, from whales to hummingbirds, from land estates to window-boxes. I can see only one such force: a respect for land as a living organism; a voluntary decency in land-use exercised by every citizen and every landowner out of a sense for and obligation to that great biota we call America."

A New Agrarianism is that decency. And as we move deeper into the twenty-first century, the issues of decency, food, hope, joy, and good land use couldn't be more important. Our health and wealth depends on what we choose to eat, how we produce our energy, where our water comes from, and who benefits from sustainable practices—and each has its root in the land.

This is the fifth wave—sustainable food production from farms and ranches that are managed for land health, biodiversity, *and* human well-being. It is a vision of New Agrarians working to sequester carbon in soils, improve water quality and quantity, restore native plant and animal populations, fix creeks, develop local energy sources, and replenish the land for people and nature alike. It is a vision of coexistence, resilience, and stewardship—a place for people in nature, not outside it.

As Severine demonstrates, this wave is being led by youth—as every wave before it has been. The difference, however, is that today's young agrarians can stand on the shoulders of their predecessors and thus see farther. I have no doubt that what they see is both energizing and daunting, but I am equally confident they have the skill sets and the right attitudes to tackle these challenges. Fortunately, the toolbox at their disposal is full of ideas and practices that have been tried and tested in the field already. Undoubtedly, they will innovate new ones to go along with what we know works. Our role is to provide as much mentoring, inspiration, and encouragement as we possibly can.

I can't wait to see what happens next.

ACKNOWLEDGEMENTS

Books, like plants, have roots—often deep ones. The roots of this book extend back to 1995, when I became active with the New Mexico Chapter of the Sierra Club and asked permission to write a regular column for the Chapter's newsletter. It was a cheeky request. Not only was I a total newbie to environmental activism, I had never written anything publicly before on conservation. I titled the column *The Uneasy Chair* (in a nod to Wallace Stegner) and I expected a polite "no" to my request. Instead, Barbara Johnson, the newsletter editor, said "you bet" and then stood bravely beside me when my musings on the environmental movement stirred a backlash among a handful of Club activists. The slings and arrows continued to fly during the early years of Quivira, but Barbara's support never wavered, so I kept writing. Ultimately, these baptismal writing efforts became the seed that would bloom into this book.

In early 2003, I approached Greg Lakes, editor of the online regional newspaper *Headwaters News*, about publishing an occasional column profiling innovative ranchers, conservationists and scientists, which I titled *A West That Works*. He agreed and for the first time my writing reached beyond New Mexico. At the same time, I began to try my hand at longer essays, including 'The Working Wilderness' which I submitted to a national conservation magazine with the endorsement of Wendell Berry, who I had met recently. Rejected by the magazine, I was honored and astonished when Wendell asked if the piece could be included in a collection of his essays titled *The Way of Ignorance* in 2005. It was an extraordinary thing to do and I remain humbled by his generosity to this day.

Buoyed by the support I was getting—and by the joy of writing—in the spring of 2006 I gathered my columns and essays together into one large stack, literally. During a sunny week in a cabin on the James Ranch, I shaped them into a book. When I was done, I sent the manuscript to Rick Knight, a professor and Quivira Coalition board member, for his feedback. He called and urged me to submit the book to Barbara Dean at Island Press, which I immediately did. To my astonishment once more, Island decided to take a leap and *Revolution on the Range* was published in May, 2008. It was a bewildering and heady sequence of events and I can't thank everyone involved enough.

The seed had become a green plant and now it began to grow.

More writing and publishing followed. In particular, I want to thank David and Elsie Kline, publishers of *Farming* magazine, and the editors at *Acres* magazine for printing many of my next round of essays. The Quivira Coalition continued to publish articles as well, giving me an opportunity to test the waters with new ideas. In early 2007, for example, I debuted the term 'Age of Consequences' as part of the title of our annual conference (without raising too many eyebrows). Then, in 2010 I tried another far-out idea—a carbon ranch—on our annual conference's audience. Its positive reception spawned an idea for a book about soil carbon which, two years later, was accepted by Ben Watson and Chelsea Green Press for publication. It felt like another leap of faith and I'm deeply thankful for everyone's support.

In the meantime, I worked quietly on the *Chronicle of the Age of Consequences* which was published online in a series of installments. I am very much indebted to webmaster Deborah Myrin for her skill and patience in making the Chronicle a success.

In 2012, it all came together during a sabbatical. At a two-week retreat in northern Wyoming, hosted by U Cross Foundation, I assembled the first draft of the *Age of Consequences*. The residency also gave me an important opportunity to contemplate a 'career' as a writer of books. It was a personal tipping point, as it turns out, and my appreciation goes out to the good folks at the U Cross for their support. Many thanks also to the staff and board of the Quivira Coalition, especially Avery Anderson who stepped up to become acting Executive Director in my absence—and took over permanently when my sabbatical ended. I also want to thank the extended Quivira family and all the innovative, inspiring and amazing people I met in the course of directing the organization for fifteen years. It was the adventure of a lifetime.

I am hugely grateful to Jack Shoemaker and Counterpoint Press for publishing this book. I am also indebted to Wendell for making introductions and blessing the project with his support. For decades, Counterpoint has been one of my favorite publishers and I am honored to be counted among its authors.

I would also like to acknowledge a few of my favorite authors, all of whom inspired me at one point or another in my life with their creativity: Joseph Conrad, Edward Abbey, William Faulkner, Wallace Stegner, Charles Dickens, John Muir, William Shakespeare, Aldo Leopold, Arthur Miller, and Wendell Berry. I inherited my love of literature from my mother, Joyce, a voracious reader and prolific letter-writer. Words meant everything to her and although she never achieved her dream of becoming a writer herself, she passed along her passion to her children. I didn't fully appreciate this inheritance at the time, but I do now.

Lastly, this book would not have been possible without my family—Gen, Sterling and Olivia. Not only is it a product of many happy trips the four of us took together over the years, now tender memories, the almost daily process of thinking about the world that Sterling and Olivia will inherit shaped many of the subjects and themes found among these pages. Creating a hopeful future was a primary motive for my work with the Quivira Coalition and as Sterling and Olivia grew older my determination to make the world a better place doubled, then tripled. Whether I did enough to make their patrimony a positive one remains to be seen, but I am forever indebted to them for inspiring my work and enriching my life. Gen too—my partner in this grand adventure. Her support and love and advice proved invaluable, over and over. Together, we explored a lot of territory, laughing, talking, holding hands and sharing the ups and downs.

I'd do everything all over again in a heartbeat.